ROGER WILLIAMS COLLEGE LIBRARY
3 1931 00220 1365

D0857873

Human Posture

The Nature of Inquiry

John A. Schumacher

State University of New York Press

Published by
State University of New York Press, Albany

Printed in the United States of America

For information, address State University of New York
Press, State University Plaza, Albany, N.Y., 12246

Library of Congress Cataloging-in-Publication Data

Schumacher, John A.
Human posture : the nature of inquiry / John A. Schumacher.
p. cm.—(SUNY series in science, technology, and society)
Bibliography: p.
Includes index.
ISBN 0-7914-0120-0.—ISBN 0-7914-0121-9 (pbk.)
1. Science—Methodology. 2. Science—Philosophy. 3. Science—Social aspects. 4. Body, Human (Philosophy) I. Title.
II. Series.
Q175.S413 1989
502.8—dc19

88-38268
CIP

10 9 8 7 6 5 4 3 2 1

Contents

Preface

It is with great anticipation, but some significant trepidation, that I write this preface, seven years after I began the initial research for this book. I am not sure I yet believe that in a few months the result will begin its own adventure in the world. As my first book, as the fruit of my life's learning to this point, it is not so easy finally to finish. Indeed, in no small way, this book *is* my life.

Yet twenty-two years ago when I began to study philosophy, I could not have imagined that I would write this book. After switching from mathematics to philosophy in graduate school, while working as a systems engineer for IBM, I would have been the last to believe that any notion of human posture would come to occupy the center of my intellectual life. Nor would I have taken myself ever to be interested in trying to question the nature of philosophical inquiry. In those days I was glad enough just to be able to find my way in philosophy as I found it.

Today, however, I would urge the readers of this book to suspend, as much as possible, their preconceptions about philosophical inquiry. I do not even wish to label what I try to do in the book, though it is easy to say that the underlying motivation is to resolve the question as to how to live. It was my own life, after all, that eventually led me, indeed, challenged me to try to create a new center for an intellectual life. I had to lose my way in life in general in order to find myself in a position to ask, as if for the very first time, just what terms I needed to use to satisfy my drive for understanding.

Many persons have helped me along the way, though, as authors say, none of them is responsible for any lack of understanding that has found its way into this book. In general I wish to thank my students and colleagues over the last eighteen years for bearing with me as I gradually worked my way out of one intellectual life and into another. My chairperson at Rensselaer, Shirley Gorenstein, provided much needed moral support and guidance, while Rensselaer itself granted me a sabbatical to support my research during the calendar year 1983. Marge McLeod, Carol Halder, and Frances Anderson, as always, provided generous and joyful secretarial support. And at the end the editor at SUNY Press, Rosalie Robertson, along with her staff and the reviewers, all assisted in crucial ways in the completion of the book.

But most crucial were my friends, in particular, for this book, Robert Anderson, Cedric Evans, John Fudjack, Sheldon Heitner, Reini Martin, Ellen McLaughlin, Mark Nasuti, Nicole Pertuiset, Sal Restivo—who is also the general editor for this series—Jean Rose, Pamela Jean Rosi, Hugh Scott, Richard Stemm, Marion Wade, Ken Warriner, Meg Wichser, and David Wieck. I want especially to thank David, who never tired throughout the whole of my intellectual life in challenging me to work through my misgivings and share the results with the world. And Hugh, whose untimely death while I was drafting the introduction and closing section of the book, reminded me of the very reasons for writing it in the first place. And Pamela Jean, whose birthday it is today, and whose generous heart allowed her to understand me in a way that I had not realized I could be understood.

Finally, then, I dedicate this book to my parents, Bella and William, who have persevered in their love for me, and to my daughter, Rachel, who, more than anyone else in my life, helped me understand how to trust another person.

April 9, 1989
Troy, New York

Introduction

I. Objectives

Bees perform dances in their hives to indicate to other bees where to fly for pollen. If, however, they are cut off from the light of the sun, their dances become disoriented. Even in their hives, bees are dependent on the light of the sun in order to be oriented in the world. Although early human beings were similarly dependent on the light of the sun, we modern human beings may well never know where the sun is. We can stand in front of a you-are-here map that orients us because of the way we face the world. Exactly as we can be oriented from our homes, we proceed by turning left or right, or by going forward, without any awareness of how these directions are correlated to the light of the sun. We are, so to speak, loose from the sun.

Erich Neumann had this to say about the advent of our looseness:

> Space only came into being when, as the Egyptian myth puts it, the god of the air, Shu, parted the sky from the earth by stepping between them. Only then, as a result of his light-creating and space-creating intervention, was there heaven above and earth below, back and front, left and right—in other words, only then was space organized with reference to an ego. (1973:108)*

But why the god of the air? What does the air have to do with left and right? Neumann did not take the Egyptian myth literally. In referring to an ego, he missed an opportunity to discover the terms of posture: what kind of place must we make in the world with our bodies if the Egyptian myth is given a literal interpretation? As it turns out, this interpretation also serves to explain how human beings first developed the ability to read silently.

---

*The dates of references are specified only if the bibliography includes more than one work by the author in question.

Today, despite our public schools, not all of us actually exercise this ability. Picture the physical arrangement of students and teachers in typical classrooms in our public schools. Following Neumann, we would no doubt interpret the arrangement with reference to an ego, that of the teacher, who alone makes a place at which he or she can meet the eyes of everyone else. Teachers face their classrooms as if they were standing at the you-are-here spot on a map. The students are automatically in the front or the back, to the left or the right, and they feel it too: no student ever forgets the difference between sitting in the front and sitting in the back of a classroom.

The physical arrangement of students and teachers is in the space created by the god of the air, Shu. The parting of the sky from the earth—above and below—cannot be undone by movement: places in space stand apart once and for all. The very first classrooms were more like the sky and the earth before they were parted. Students and teachers were interwoven in a pattern of movement. Their physical arrangement was, in other words, dynamic, certainly incompatible with the modern classroom seating chart: from the perspective of the you-are-here spot of the teacher, the places of the students stand apart once and for all. Even more profoundly than the parting of seats, the parting of classes—what we call "age-grading"—cannot be undone by movement. What could a student be thinking of if he or she were to decide simply to move to a seat in the classroom of another grade? Yet not so long ago a student could have thought: "I *can do* the work."

When I tell people that I have written a book about human posture they invariably express interest and curiosity. We modern human beings already believe posture to be *an* important aspect of life, though not, as I seek to show, *the most* important one. We know that we make places in the world with our bodies, but we tend to confine our concern about these places to whether or not, for example, we slouch in our chairs. We are not at all clear about the abilities of our own bodies, even in routine matters of physical health. Most of us do not even believe that we are bodies in the end anyway: we do not accompany our bodies to *their* graves.

We are also not clear about exactly how our bodies are affected by the process of socialization in our society, for example, the cumulative effect of the physical arrangement of students and teachers in typical classrooms in our public schools. Is the problem of illiteracy today (Kozol) a problem of our posture? Or again, can a literal interpretation of the Egyptian myth serve to explain not only the origin, but also the subsequent development, of our ability to read silently? Together we tend to make places in patterns, and in terms of these patterns we can account for how we think and feel.

To call our everyday beliefs about bodies into question in the book, especially as I do in part two, by reconsidering the history of physics from Isaac Newton to today, my life as a philosopher in a university is just as relevant as my everyday life in our society. Throughout the book I am interested in applying the discussion as much to the matters of my profession as to everyday matters. It is this double concern that makes this introduction so hard to write without employing the terms of posture themselves: only through them do the professional matters become just as transparent as do the everyday matters. I cannot draw examples from part two as readily as I can from part one—the example of the bees—and part three—the example of the classroom, to which we will also return in section two.

The task of this introduction is to state the objectives of the book without employing the terms of posture themselves. In some instances, these terms require extensive development before they can be employed, but more fundamentally, in many instances, their force depends on their very transparency once they are put to work. The method of the book, especially in part one, is gradually to build a certain suspense in the development of individual inquiries. Then the terms of posture can offer an immediately obvious resolution. Indeed, although their own development is just that of our language in general—I will draw them out, as a kind of genealogy, from the roots of quite familiar terms—it is as if they were actually designed for their work here. One of the objectives of the book is to display this transparency.

In professional circles, my quick explanation of what I am about in the book is that I aim to be, as Richard Rorty put it (360), "edifying": to find "new, better, more interesting, more fruitful ways of speaking." We engage in "the 'poetic' activity of thinking up such new aims, new words, or new disciplines" so that we can "reinterpret our familiar surroundings in the unfamiliar terms of our new inventions," taking us "out of our old selves by the power of strangeness." We aim to improve our ability to cope with our lives and the world. For the purposes of edification, at any rate, "the way things are said is more important than the possession of truth" (359). Or again, we try "to keep the conversation going rather than to find objective truth" (377), especially if we think of objectivity as we have done at least since the dawn of the seventeenth century, a topic to which we will return in section three.

No doubt the terms of posture *are* edifying. As I wrote this book, I was surprised over and over again just how extensively these terms can be used to reinterpret our lives and the world. About seven years ago I had an intuition that I had finally found the ground on which I could pull together my work to date into a whole philosophy. Influenced by Michel Foucault's and David Bohm's works on the body, but just as much by the challenge of my own everyday life in a world I no longer understood well enough, I came to believe that the body was somehow the key to unraveling the mysteries of how to cope with our lives and the world. Little did I know how much I would be taken out of my old self by "the power of strangeness" of the terms of posture.

What started out, and still may be seen, as a work of edification became to me much more. To place this added dimension in the context of my profession will take the entire first two parts of the book, though I do not stop developing this context thereafter, inasmuch as part three extends the range of inquiry in question from the natural to the social sciences. At the end of part two, I finally offer a criticism of Rorty's own discourse that reveals a new sense of objectivity, what I call "the order of co-making of inquiry": we do not so much construct the world socially through our discourse as we work together with the world to construct the whole show, *including* our discourse. Only in the terms of posture, moreover, can we develop this new sense of objectivity. It is this claim of the book that makes it more than edifying, indeed, even aiming to be foundational in the end.

Again, I set out to write a book about the entire range of human inquiry—the disciplines in which we *raise questions* about our lives and the world—without recourse to any other terms than those of posture. If I could accomplish this task, I thought, the sheer scope of the final result would in itself constitute sufficient defense of the exercise. As constituting one discourse among others, the terms of posture could possess "the power of strangeness" by displacing our usual reliance on the categories of mind or consciousness that have been considered to transcend body: hence the edification, which could not always be concerned with fitting into the historical context of more traditional techniques of philosophical analysis. But, beyond the edification, the discourse of the terms of posture actually turned out to enable me to develop—continuing the example from part two—a new theory of how our nervous system works. On this theory, we do not reduce the discourse of mind or consciousness to that of body; we eliminate mind or consciousness exactly as Einstein eliminated the ether in physical theory. The discourse of the terms of posture provides some *answers* as well.

Perhaps the best term I have found for the way in which anything could be an answer in the new sense of objectivity is the Greek term 'mythos', which, according to Giambattista Vico, "came to be defined for us as *vera narratio*, or true speech" (1984:§401). As I show in part one, it is best to think of the original context in which we began our inquiries into our lives and the world as one in which both the questions and answers were taken to be *between* us and the world. Along the way to the modern world, we ruptured this connection to such an extent that both the questions and answers were no longer in the world at all, but rather *in us*. Rorty rightly criticized these attempts to take mind or consciousness to be "the mirror of nature." But his pragmatic commendation of "new ways of speaking," even with "the power of strangeness," is not enough either: I aim here to convince the reader that the terms of posture constitute a mythos in its original sense of "true speech" between us and the world.

Moreover—and here again I felt the influence of Foucault, in particular his 1977 interview "Truth and Power" (1980:109–33)—the "true speech" of the terms of posture turned out to establish "the possibility of constituting a new politics of truth" (133) in such a way that I finally decided that the speech had been purposely dismantled to the extent that it once existed and otherwise had been simply repressed. It is especially clear in the terms of posture that the present "discursive regime" "*governs* statements" (112–3) so that the ground on which we could trust each other to work together to *re-make* our society—the ground on which, as Vico also put it, we *made* our society in the first place—cannot be evident to us. Indeed, the "forms of constraint" (131) that produce what currently passes as truth for us are even more open to Foucault's political criticism than they are to Rorty's epistemological criticism. I aim here to convince the reader that the terms of posture constitute a "true speech" of *personal* liberation, throughout everyday life.

II. Aleatory Body: Personal Liberation

The word 'aleatory' is derived from the Latin word for a die, which we normally think of in pairs, as dice. What is aleatory depends upon "the throw of a die," upon "uncertain contingencies" (*Oxford English Dictionary*, hereafter, *OED*)—it is a matter of chance. A die has six faces, usually differentiated by the number of dots on each face, from one to six. We play games in which we throw a die and bet on which face will turn up when the die stops moving. We know that, other things equal, each face has the same chance of turning up, though we also imagine that somehow we can anticipate the uncertain contingencies that govern a particular throw of a die. We may even go so far as to imagine, for example, that, if one throw turns up one face, the next throw will be less likely to turn up that face.

In any event, we also know that the uncertain contingencies of the throw of a die have nothing whatsoever to do with the dots on each face. Or again, as far as the movement of the die is concerned it does not matter how the faces are differentiated, if they are differentiated at all. In the betting game, however, the differentiation of faces is the crucial matter: we can change the odds of the betting game by changing the dots on the faces, for example, by putting an extra dot on the face with one dot now—we can even imagine changing the game while the die is in flight—though this change could not affect the order of movement of the die itself. Let us call what we do to recognize the faces of a die "reading the die," so that we can say that the movement of a die and its reading are independent of each other, except in the sense that, once the betting game is given, we have no choice but to read the die as it determines by coming to rest. The die turns up a face in the context of a certain betting game, but *only in that context* can we read the face to discover which bet anticipated it.

If we take a die as the paradigm of an aleatory body, then an aleatory body must be open to two kinds of description: one in which it is an order of movement alone, and another in which it is read. Take any body, throw it, and bet on which of its faces, so to speak, will turn up when it stops moving. Each kind of body, indeed, each particular body of a certain kind, will be governed by the set of the uncertain contingencies of its movement, but unlike a die the set will undoubtedly favor, even if other things are equal, one face of the body over another. The faces themselves may actually be differentiated at the level of the movement of the body, though this differentiation is not a matter of reading. As far as reading the body is concerned, we must imagine that each of its faces can be treated as if it were to sport just so many dots, dots that are independent of the order of movement of the body itself. They would merely be aspects of the betting game that constitutes the context of reading in which we imagine throwing the body.

Let us now imagine a particular betting game in which the body is a human doll, in itself not obviously male or female. We are going to throw the doll, and when it stops moving it will turn up in some spatial arrangement of its parts relative to each other. Many such arrangements are what we typically think of as postures: were a doll-sized chair in the right place, for example, the doll might end up slouching in it. Let us imagine then, still working within the limits of the typical sense of posture, that we take bets on whether the final arrangement of the parts of the doll to each other will be read by us as a male posture or as a female posture (or as no posture at all in most cases, I would guess). That the gender of a human doll can be "read at a flash" in this way is not in doubt (Goffman, 1979:27). Moreover, although the arrangement of the various parts of the doll to each other is not independent of its movement, the order of movement of the doll cannot in itself determine how we may read it. The context of reading that this game constitutes can vary from culture to culture; quite often what one culture reads as female another reads as male, and vice versa.

A human doll is an aleatory body, open to two kinds of description: one in which it is an order of movement alone, and another in which it is read. At the limits of the typical sense of posture, the possible readings become more sophisticated, as illustrated in this sentence from the *OED*: "the artist draws her in three distinct postures, like a captive, like a penitent, and like a conqueror." We can extend the above imaginary game to include such readings—of social roles in general—though as we do so the game itself will have to become more sophisticated, if only to include some props that we imagine throwing along with the doll, clothes for example. (We could imagine throwing clothes along with the doll as if the doll were one die, the clothes another, and we were betting on the final arrangement of the two.) In other directions, we may include gestures (Morris *et al*), and allow the end result of our throwing the doll to remain a movement of a certain kind, as if the doll were an automaton perhaps. As we build up the game we come closer and closer to simulating our social life itself. (We could imagine throwing many dolls and props as if each were a die and we were betting on their final arrangement.) As Erving Goffman put it (1979:27), here we "discern how mutually present bodies, along with nonhuman materials, can be shaped into expression."

Always keep in mind, however, that, no matter how sophisticated the game becomes in terms of the readings of doll, these readings are never determined by the order of movement of the doll alone. At the level of movement, the doll *can* move in certain ways, ways that depend solely on the spatial arrangement of the parts of the doll to each other. I wish to base my use of the word 'person' on this description of the doll as an aleatory body, so that it can be said that readings of a person do not constitute that person: a person is constituted by what one *can do* at the level of movement alone. But now we must change the nature of our imaginary game: instead of imagining that we throw a human doll, or rather, a doll of a person, we imagine that persons throw themselves.

Here we introduce a host of related issues that turn on the voluntary character of the movement of a person as opposed to that of a die or a doll. The very distinction between moving and reading is called into question in the sense in which a person can move, indeed, *should* move so as to be read in a certain way. If a person is aware neither of an option to move so as to be read in an alternative way, nor of an option to move in the same way but to be read in an alternative way, one acts as if the way one moves were to determine the way one is read and vice versa. Given such a lack of awareness, what would otherwise be an uncertain contingency becomes as much of a certainty as we can have in the "betting game" of our social life: the context of reading seemingly comes right out of our bodies, not really a matter of a game at all. Hence it is possible to become alienated from oneself as a person or, equivalently, from oneself as an aleatory body. The way one lives comes to be constituted by the way one is read.

Within the game of our social life "it is as if perception can only form and follow where there is social organization" (Goffman, 1972:382). The set of permissible readings of persons must be internalized in such a way that we tend to become alienated from ourselves as persons:

> The self is a code that makes sense out of almost all the individual's [in our terms, the person's] activities and provides a basis for organizing them. This self is what can be read about the individual by interpreting the place he takes in an organization of social activity as confirmed by his expressive behavior [in our terms, his movement]. The individual's failure to encode through deeds and expressive cues, a *workable* definition of himself, one

> which closely enmeshed others can accord him through the regard they show his person, is to block and trip up and threaten them in almost every movement that they make. The selves that had been the reciprocals of his are undermined. (366)

Should we step out of the certainties of social organization into the uncertain contingencies of movement as an aleatory body, we risk what Goffman called "insanity of place." (One of my favorite ways of illustrating insanity of place, again drawn from Goffman, is to ask people to imagine what it would feel like to approach someone else's cart in a grocery store and then, without uttering a word, to take an item out of it.) Jacques Donzelot, Richard Sennett, and Foucault showed how, as Goffman put it (187), we "bear the cross of personal character in the presence of others": we read a person for one's "self," to determine what a person *is*, not what a person *can do*. Precious few, if any, aleatory openings exist in our socialization; we arrive on the adult scene without sufficient sense that we *can* move or be read in alternative ways. Sennett thought of this as a loss of a sense of *play* in social life—hence the aptness of the metaphor of alternative "betting games."

The persons in a typical classroom who are read as "students" or "teachers" may not be aware of themselves apart from their readings, that is, on the level of their *all* being thrown in the first place. Nevertheless, we can easily imagine a betting game in which all the persons in a classroom are thrown, and one possible, perhaps even likely, outcome is that we read one person as "teacher" and the others as "students." This reading has simply become a certainty of the social situation of a typical classroom. It is as if the person who plays the teacher were not thrown at all, and as if the persons who play the students were thrown by the teacher: the teacher throws the students, no doubt along with some paper and pencils, perhaps even establishing odds for the end results, and then reads those results by grading them, as if the grades were resolutions of dots on the students' faces. Indeed, it is *only* through the teacher's reading of the students that they can move on to the next level of schooling, whether it be for the next hour or for the next year. The actual movements of the person who plays a student cannot by themselves determine his or her reading and in turn advancement, nor of course can the movements of the person who plays a teacher determine his or her reading.

At the level of aleatory body/person, in other words, the throw of *all* the persons present allows some chance that *any* one of them *can* do whatever is required (or that *any* one of them *cannot* do whatever is required). Put yet another way, at the level of aleatory body/person the roles of student/teacher are *reversible*. To act in an aleatory way in a typical classroom, therefore, is to risk insanity of place. Indeed, we would be hard put to find *any* social situation in which acting in an aleatory way would not incur that risk. Think of the social roles in the home: child/parent-or-adult (or wife/husband, mother/

father). Or in the work place: employer/employee. Or in society at large: citizen/government official, such as a policeman. Nowhere can we act as if things were still up for grabs, as if we were *all* thrown in the first place. We can easily imagine cases, even amusing cases, of returning to the level of aleatory body, such as an aleatory street on which we must rediscover every day what readings apply to the direction of the traffic (no doubt we would pay much more attention to the order of movement of the cars around us). But is returning to the level of aleatory body practical as well?

George Dennison advocated, in effect, that we must realize the level of aleatory body in order to overcome such educational problems as that of illiteracy. At the extreme of a typical classroom—an extreme far too many of our young persons experience (Kozol)—a person may come to feel as if the very words before one's eyes were always already possessed by the teacher. Were one to throw oneself into it, as we say, one would never turn out to be able to do whatever is required; only the teacher could do it. Dennison urged us instead to "bring the bodies back":

> I wanted to get back, in some way, to the stage of reading at which written words still possessed the power of speech. And so our base of operations was our own relationship; and since José early came to trust me, I was able to do something which, simple as it may sound, was of the utmost importance: I made the real, the deeper base of our relationship a matter of physical contact. I could put my arm around his shoulders, or hold his arm, or sit close to him so that our bodies touched, or lean over the page so that our heads almost touched. Adults, and especially adult Americans, are not used to this kind of touching . . . [but] the importance of this contact to a child experiencing problems with reading can hardly be overestimated. . . . Nor was it my body beside his that meant so much, but the fact that the presence of my body vivified his awareness of his. He knew where he was: he was in his skin; and when little bursts of panic made his head swim or his eyes turn glassy, he did not have to run away or reject the task *in toto*. He could gather himself together, because his real base—his body—was still there. (168–9)

I would only add that one's real base is an aleatory body: we are bringing the bodies back as aleatory bodies, rather than as always already read to be students or teachers. *Let us bring the persons back.*

Dennison threw himself *together* with José and the other young persons in his experimental First Street School. His own movement was, as much as he could make it at least, a matter of uncertain contingencies quite like those of the movements of the persons with whom he wished to share his abilities. They all felt the level at which the social roles of student/teacher are reversible:

> When adults stand out of the way so children can develop among themselves the full riches of their natural relationships, their effect on one another is positively curative. Children's opportunities for doing this are appallingly

> rare. The school life is dominated by adults, and after school there is no place [that is, no aleatory place] to go. The streets, again, are dominated by adults. . . . Perhaps the most important thing we offered the children at First Street was hours and hours of *un*supervised play. . . . Indeed, on several occasions with the older boys, I averted violence simply by stepping out of the gymnasium! (82–3)

Once the reading of a person as "teacher" is a certainty of the social situation of a classroom, the persons who are read as "students" must learn, on their side, how to move so as to gain an advantage over their fellows in the eyes of the teacher; as students, persons are isolated from each other in a competition not of their own making. Unless a person enters the competition as "student," one is constantly aware of the price one is paying: to throw oneself into a typical classroom is necessarily to foresake the full riches of one's natural relationships as a person to other persons. The same can of course be said of persons who throw themselves into it to become teachers. Hence the only way to break the spell of the readings of a typical classroom—that is, throwing oneself *together* with others—appears to everyone as the most heretical option of all.

To break the spell of the readings in any context, if only to know exactly what these readings are, I cannot read another's movements: I must *move with* that person so that the outcome of our *joint movement* is an uncertain contingency, as if each of us were a die and we were to throw ourselves together. We realize the level of *what we can do together*. And even more important, we can only find out in the end what we can do. We have, at least at first, no readings to form and guide our perceptions. We must perform, so to speak, *a dance improvisation* in which we pay *direct* attention to each other's movements. Here is a perfect example—what I take to be an aleatory gymnasium experience—from *The Peckham Experiment*:

> The boy who swings from rope to horse, leaping back again to the swinging rope, is learning by his eyes, muscles, joints and by every sense organ he has, to judge, to estimate, to *know*. The other twenty-nine boys and girls in the gymnasium are all as active as he, some of them in his immediate vicinity. But as he swings he does not *avoid*. He swings *where there is space*—a very important distinction—and in doing so he threads his way among his twenty-nine fellows. Using all his facilities, he is aware of the total situation in that gymnasium—of his own swinging and of his fellows' actions. He does not shout to the others to stop, to wait or move from him—not that there is silence, for running conversations across the hall are kept up as he speeds through the air.
>
> But this "education" in the live use of all his senses can only come if his twenty-nine fellows are also free and active. If the room were cleared and twenty-nine boys sat at the side silent while he swung, we should in effect

> be saying to him—to his legs, body, eyes—"You give all your attention to swinging; we'll keep the rest of the world away"—in fact—"Be as egotistical as you like." By so reducing the diversity in the environment we should be preventing his learning to apprehend and to move in a complex situation [or rather, in an aleatory situation]. We should in effect be saying—"Only this and this do; you can't be expected to do more." Is it any wonder that he comes to behave as though it is all he *can* do? By the existing methods of teaching we are in fact inducing the child's *inco-ordination* in society. (Pearse and Crocker, 192)

On the contrary, we are inducing the young person to dance to the rhythm of his readings in society, especially that of being "a child" in the first place.

III. Tasks

When I said in section one that I was influenced by the challenge of my own everyday life in a world I no longer understood well enough, I had in mind primarily the very end of section two: the inducing of young people to dance to the rhythm of their readings in society. I had lost my rhythm and, as a result, felt as if I were dangerously amorphous. At the time I had not yet formed the conception of aleatory body/person, nor had I discovered the terms of posture, the terms in which the entirety of section two takes on its proper force. Influenced largely by the work of Philippe Aries (1962), I still realized right away that the institution of childhood—including all that came along with it, especially compulsory schooling—had not been designed to lead me down the path to my liberation.

Again, if a person is aware neither of an option to move so as to be read in an alternative way, nor of an option to move in the same way but to be read in an alternative way, one acts as if the way one moves were to determine the way one is read and vice versa. One is alienated from oneself as an aleatory body/person. The institution of childhood, more than any other, furthers such alienation. Because Dennison was confronting the very foundations of our society, his method carries force well beyond its intended scope. Only in the terms of posture, however, does its full force become evident. For reasons that emerge in part three, I call working at the level at which we can throw ourselves *together*—the level at which we *all* must be thrown in *joint movement*—"the osmosis method." Although this method is effective in all areas of everyday life, it is especially effective when one has lost one's rhythm; it counteracts the tendency to turn back to one's readings in society, which constituted one's problem in the first place.

Even though my own practice of the osmosis method is confined largely to young persons at a university who are considerably older than Dennison's young persons, they still embody the same alienation: they have been led to live, from birth, as if they were throwing themselves at their parent's feet, as we say, to be read as acceptable or not. They hardly notice, if they notice at all, that their parent, like their teacher, does not throw him or her self before them, let alone at their feet. Social roles such as these constitute one-way streets on which no aleatory bodies/persons are allowed to travel. As Sennett claimed, we no longer have a sense of *making*, let alone *co-making*, our social life: it is always already choreographed. All we can do is learn how to dance in the way that it prescribes; we should not even realize that this is what we are doing. Hence we make society into an other, *them*, not us: I will dance to the rhythm of a reading of myself, indeed, a reading *they* give to me, not even one I give to myself—I will practice what in an earlier work (1961) Goffman called "the art of shamelessness."

As difficult as it will be in most of our cases to realize the aleatory ground of full personhood, the osmosis method offers us the only chance of really approaching that ideal. Especially in our families and schools, we must work as much as possible on the level at which it makes sense to bet as to whether or not *all* the persons involved *can* do whatever is required. We must develop the ability not to respond to the level of description of ourselves that is called "reading," so that we can begin to ask—or rather, to re-ask—the questions behind the particular readings our society offers as answers. Perhaps surprisingly, the more private social life becomes the more difficult it is to learn to ask such questions. Although the family is often thought of as a refuge from public role pressure, to move as if the social roles of child/parent were reversible is to risk the most severe insanity of place. Fortunately the osmosis method calls upon the person who is read as the child/parent just to realize the level at which one throws oneself into one's family in the first place, the ground on which it is *possible* to reverse the child/parent roles.

The unique force of the osmosis method is that it produces results even if one fails at it: one need only *try*. It will take the entire book to create the context in which this claim will be self-evident, though by the end of part one we will understand how the osmosis method taps the residue of our original posture, a posture in which we understood joint movement to be the very nature of the world. The special task of part one is to characterize the advent of the posture—the modern human posture—in which we came to read the world. It was this posture that gradually seduced us to abandon joint movement as a way of life, and instead to dance to the rhythm of our readings in society. Only in our original posture is the co-making of inquiry self-evident. Inquiry in the modern human posture, though responsible for many advantages of our current way of life that we cannot afford to lose, does not make co-making evident. It will take both parts two and three to establish the context in which even reading the world *is* an order of co-making of inquiry.

Perhaps the best way to introduce the problem of objectivity that we confront in part two—the one that Rorty had to confront as well—is to listen to Galileo Galilei characterize the enterprise of inquiry which he took himself to be conducting during the early decades of the seventeenth century:

> Philosophy is written in that great book which ever lies before our eyes—I mean the universe—but we cannot understand it if we do not first learn the language and grasp the symbols, in which it is written. This book is written in the mathematical language, and the symbols are triangles, circles, and other geometrical figures, without whose help it is impossible to comprehend a single word of it; without which one wanders in vain through a dark labyrinth. (Burtt, 75)

Arriving at objective truth is like successfully reading a book, a book always already written by God, just waiting for us to discover how to read it, as if it were open to one and only one correct reading. Evidently no co-making of inquiry between us and the world is relevant here. We are broken apart from the world, and indeed the nature of the sensations that constitute our experience of the world is not the nature of the world itself:

> But that external bodies, to excite in us these tastes, these odours, and these sounds, demanded other than size, figure, number, and slow or rapid motion, I do not believe; and I judge that, if the ears, the tongue, and the nostrils were taken away, the figure, the numbers, and the motions would indeed remain, but not the odours nor the tastes nor the sounds, which, without the living animal, I do not believe are anything else than names, just as tickling is precisely nothing but a name if the armpit and the nasal membranes be removed. . . . (Burtt, 88)

In the last decades of the seventeenth century, Newton would claim to show exactly how, if the eyes were taken away, no colors would remain: colors are *in us*, entirely broken apart from the world of bodies, indeed, not even in our own bodies.

The special task of part two is to employ the terms of posture to reinterpret the difference between the physics of Newton—what he called "natural philosophy"—and the current physics of relativity theory and quantum mechanics. This difference is interesting in itself: even before we reinterpret it in the terms of posture, we must realize the limitations of conducting inquiry by reading the world. Once reinterpreted, however, the difference also constitutes the foundation for a new theory of how our nervous system works that reconnects us to each other and the world *through our very bodies*. Not even reading the world, let alone our experience of colors, can be broken apart from the world. Reading too is *between* us and the world, just as much an order of co-making of inquiry as is our original mode of inquiry, that of joint movement, which we can conduct again through the osmosis method.

This result allows us to develop, as the initial task of part three, the concept of a person introduced above. In section two, we were forced to rely on the order of movement of body because we had not yet developed the terms of posture themselves. The crucial aspect of the order of movement of body is that of *joint* movement, but after part two we realize that the order of reading of body is one of *joint* reading as well. Hence the technique employed in section two—or rather, the metaphor of the "betting game"—cannot in itself provide the ground for understanding the extent of our being thrown *together* in society, indeed, in the world. Whether in professional or in everyday affairs, the new objectivity and its associated concept of a person constitute a new community.

By the time Thomas Kuhn wrote the postscript to the enlarged edition of his well-known book, *The Structure of Scientific Revolutions*, new research in sociology led him to conclude: "If this book were being rewritten, it would therefore open with a discussion of the community structure of science" (176). The sociological research in question aims to understand science in terms of the social construction of the world, a construction that the order of co-making of inquiry attributes as much to the world as to us: *everything* was thrown *together* in the first place. Hence the result of part two allows us to develop, as the final task of part three, the professional discourse that includes Kuhn's work in the history of science as well as Foucault's and Rorty's work in the philosophy of knowledge. In the process, we will compare some of Ludwig Wittgenstein's work in the philosophy of language to Albert Einstein's resolution of the problem of "attaching meaning to statements" at the foundations of physics (22).

The current community structure of science, however, is also embedded in a certain community structure of society. It is ultimately the fundamental structure of current society that parts two and three call into question. But so far the way it is called into question has been described only negatively: we must develop the ability not to dance to the rhythm of our readings in society. In developing this ability we reach the aleatory ground on which social roles such as parent/child and teacher/student are reversible—or again, the ground on which what we can call "aleatory inquiry" is possible. But given the account in part three of the construction of the current readings of our society, a positive option can be described as well: if the rhythm of our readings in society were to change to support personal liberation, dancing to it would be another matter altogether. The objective of the book dearest to me is precisely the basic rhythm of a new dance, a dance of persons in joint movement *and* reading. It is only through *joint readings* that we can develop alternatives to our current social roles.

We *can* live in a different way. This book is ultimately dedicated to explaining why it is not, so to speak, wired into our bodies to live the way we do, especially when it comes to resolving the thorniest of our current problems, that of trusting each other enough so that we realize our ability to work together. Our solidarity lies buried in our bodies, awaiting only another physical arrangement of society to reveal itself again. Then, on its basis, we can learn how to read each other and the world in a way that promotes our mutual well-being *with* the world. If, on the other hand, we continue to fight over how to read each other as if reading were, as we say, the name of the game, we will most certainly remain at odds with each other and continue to undermine the ecology of the world. In the face of what far too many of us regard as an almost certain apocalypse, the terms of posture make it possible for us to have faith in ourselves and hope for a peaceful future.

Part One

The Origin of Inquiry

I. Posture

As I myself did at the very beginning of my research for this book, let us commence the development of the terms of posture with the first three senses of the term 'posture' in the *OED*:

(i) relative disposition of the various parts of anything, especially the position and carriage of the limbs and the body as a whole; attitude, pose;

(ii) position of the one thing (or person) relative to another; position; situation; and

(iii) a state of being; a condition or situation in relation to circumstances.

In 1695 (ii) appears in the phrase 'the Earth's posture to the Sun.' Although (ii) has become obsolete, it is still evident in (iii), the paradigm of which is the phrase 'the posture of affairs.' We can easily see (ii) in a more specific example of (iii), a phrase of 1793, 'the best posture to receive the storm.' We can also see (ii) in this example of (i), a phrase of 1633, 'the usual form of their posture at the Table.' Clearly (i) is typically thought of today, and it has a wide range of application, from 'posture at the Table' to this sentence of 1711, 'the artist draws her in three distinct postures, like a captive, like a penitent, and like a conqueror.'

The root of such terms as 'posture,' 'position,' 'pose,' and 'situation' is to-place. According to the *OED* the term 'place' begins with the sense of 'place' in this sentence of 1571, 'give place; let the prisoner by; give place,' and again in this sentence of 1602, 'for performing the play, the beholders cast themselves in a ring, which they call Making a Place.' Although the sense of 'place' in the first sentence is not as active as the sense in the second sentence, it is certainly pervasive. Let us say, a thing makes a place at all times, but the place will also vary from time to time, especially in different circumstances.

The root of the term 'circumstance' is to-stand-around. Posture is the way a thing makes a place relative to itself or to its parts, and to that which stands around it. Let us use the term 'world' to circumscribe a thing and all that stands around it. *Posture is the way a thing makes a place in the world.*

How basic is posture? Well, how indeed can a thing be *in* the world unless it makes a place there? To be in the world *is* to have a posture. Think for a moment about the different things in the world and the way they make places there. Each thing will have parts or features that are placed relative to each other in a certain order. Some relative placings are necessary, some are not, and of the latter some are possible and some are not. Let us say, each thing has both a characteristic posture that distinguishes it from other kinds of things and a range of possible, optional postures that may also distinguish it from others of its own kind. But let us also say, each thing has a characteristic posture in the first place only through its incarnation.

The root of the term 'incarnate' is to-be-in-flesh, and the term has come to cover embodiment in general, including the sense to-have-form, to-be-real. For example, in his phenomenology of the body, Maurice Merleau-Ponty spoke of a person's body as the system of all that person's holds on the world. Not surprisingly, the root of the term 'system' is to-place-or-set-together. To be incarnate, to have a body, constitutes the order of relations of making a place in the world. A thing must take a hold on the world, and this taking hold *is* its incarnation, its having a body, or its having a posture.

As a thing makes a place in the world, the rest of the world must give place to it. To have a posture a thing must be open to holds on it by the rest of the world. They are connected to each other. The root of the term 'connect' is to-bind-together. Through its posture a thing and that which stands around it are bound together. Exploring the sense of the term 'hold' we can say that a thing must take a grip on the world, seize its place there, and this gripping or seizing may well be called into question. It is in answer to this question that a thing assumes its characteristic posture, constituting it as the very thing it is.

Let me speak of myself. My incarnation places special demands on the place I make in the world, on the way the world must give place to me. I have a certain size and shape, and although I am flexible to some extent, the adjustments that I can make are limited. These limits concern such differences as that between my two-legged incarnation and another thing's four-legged incarnation. Each incarnation constitutes an order of relations of the body and its parts, or let us say, a matrix of relative places for those parts and the body as a whole. I can assume a sitting posture, whereas others cannot.

Here is the ground on which Merleau-Ponty based the significance of the "I can." Even while I am not sitting, my incarnation includes the ability to sit. It is *always* an aspect of my making a place in the world that I *can* sit. Imagine a kind of torture in which the world is arranged so as never to give place to sitting. My holds on the world must work within the matrix of my incarnation, but a significant number of the elements of this matrix will at any one moment remain options for me to display. The root of the term 'display' is to-fold-apart. An element that is not now displayed is folded together in my incarnation, always ready to be folded apart on another occasion.

To take another example, I *can* stand. I stand on my feet, upright, as we say. Imagine a kind of torture in which the world is arranged so as never to give place to standing, especially in the sense in which the term 'stand' is used in the root of the term 'circumstance.' In this sense I *always* stand around whatever stands around me. Far from being obsolete, sense (ii) of the term 'posture' is alive and well in (i): hence the bound-togetherness of a thing and that which stands around it in the order of posture.

To take yet another example, I *can* eat, but to exercise this ability food must stand around me in such a way that I *can* eat *it*. This aspect of my incarnation, of my making a place in the world, is as crucial as these aspects can be. I must take a hold on the world, at least from time to time, with my mouth. The way my mouth makes a place in the world also distinguishes me from other kinds of things, from those that are alive but have no mouth, such as an amoeba, and from those that are not alive at all, such as a rock. We should surely say as well that the way my mouth makes a place in the world distinguishes me from others of my own kind. A part of a body such as a mouth enters into both the characteristic posture and the range of possible, optional postures.

Now, given these general parameters of the order of posture, let us turn to its primitive structure. Consider eating again. I must take inside my place what is originally outside my place. After I digest it I must pass the remains from inside to outside my place. Hereby some of what stands around me moves through me, from outside my place to inside my place and then outside again, though obviously not as it was originally. If I make a place in the world, I occupy—another sense of the term 'hold'—that place: I am inside it, and the rest of the world, including my food, stands around me, outside my place. Or again, in making a place in the world, I cast everything else outside my place, as the world gives place to me, though from time to time I must allow certain things originally outside me to pass through me. Other things I must not allow inside me at all, let alone to pass through me.

So I have an inside and an outside, an exterior or surface: my skin. At my skin I meet the outside of the rest of the world. We share a boundary, at which we touch together, in contact. (The root of the term 'contact' is to-touch-together.) I am inside my place but I am outside the place of the rest of the world. All things maintain an inside/outside boundary, though unlike me not all things must allow other things to cross that boundary. Almost always, what crosses my inside/outside boundary does not begin its journey at that boundary, at my skin. It begins at a distance, standing apart from me. (The root of the term 'distance' is to-stand-apart.) Some things stand around at a distance, apart, and some things stand around at no distance, not apart, but in contact, together. If I cannot bring certain members of the former group into the latter group, I will eventually cease to make a place in the world. A certain inside/outside balance is crucial to me.

The primitive structure of the order of posture is therefore constituted by two differences: that of inside/outside, and that of together/apart or in-contact/at-a-distance. What plays the role relative to the latter that a skin plays relative to the former? A face! A face takes a hold on that which stands apart in the world. First I turn my face toward food; I face it at a distance, apart, and then at no distance, in contact, together. It is no wonder that a mouth is in a face, bound together there with an impressive set of parts or features, each of which works at a distance: typically, ears (at the boundary of the face), eyes, and nose, or their equivalents. Without an ability to take a hold on that which stands apart, a thing is profoundly confined to the place it makes, though with a skin it does have the ability to touch the outside of the rest of the world, thereby allowing it to be alive. An amoeba can be said to have a skin, a sensitive exterior or surface, but not to have a mouth: it grows around what it takes inside its place. A rock cannot even be said to have skin, let alone a mouth. Certainly the difference skin/face, perhaps even the difference together/apart, is not at work in the posture of a rock.

When it comes to such things as a flower and an amoeba, however, we may not wish to say the same thing. Should they be said to face this or that at a distance? Obviously not in the literal sense of 'to face,' though anything with a sensitive exterior can act as it if were facing that which stands apart in the world. Barring other stimulation, a human being can judge relative position and distance through being touched, essentially on the basis of the time differential between two stimulations. In their own ways both the flower and the amoeba must possess a similar ability. Nevertheless in claiming that a thing touches at a distance one runs a risk not at all involved in claiming that a thing smells, hears, or sees at a distance (though, it must be admitted, in smelling, hearing, and perhaps seeing as well, time differentials between two stimulations are essential to judging relative position and distance). I take it that this difference in risk must be traced to the difference skin/face, allowing us to distinguish between a life without a face and a life with one. Given the former life a thing needs to be, as it were, *in its food* in a way that anything with a true face never needs to be.

The difference skin/face raises a most interesting question about posture.

As we have already noted, through its posture a thing and that which stands around it are bound together. How shall we understand this bound-togetherness? Certainly it is a lot more obvious with respect to a thing that is confined to its place, without a face. Think of a flower, *rooted* in that which stands around it. So what about a thing that has a face? In what way is such a thing still bound together with that which stands around it? Don't things with faces seem to be loose in exactly the respect that things without faces seem to be bound together? Think of the bee that visits the flower. In whatever way the bee makes a place relative to the flower, mustn't we say that the bee is loose in exactly the respect that the flower is not? Are the flower and the bee loosely bound together, or is the bee simply loose, not bound together with the flower? On the latter alternative, however, won't we have to adjust what we said about posture and bound-togetherness?

And if these questions are not enough, what of a human face? If a bee is loose, certainly we are even more so. No face works as well as the human face does with respect to that which stands apart in the world, not to mention our ability, as we say, to face the future. (For Jean-Paul Sartre—and we will be better prepared to understand this in section seven of part one—a human being *always* faces the future.) Here is the key to the apparent looseness of a thing with a face: such a thing is at least loose enough with respect to its immediate circumstances to turn its face beyond those circumstances, toward what stands *apart in the world now.* Certainly occasions of such looseness dot the evolutionary trail to human beings.

What of the water creature who first ventured onto land? Or, whose first adventure was on the land? The root of the term 'adventure' is to-come-to, suggesting that what is come to does not stand together with what is left behind. Each evolutionary step is an adventure, a loosening of one thing from its immediate circumstances in such a way that it comes to that which stands apart in the world. Hence the danger or risk associated with an adventure, but also the hint of purposiveness. Mustn't we say that the water creature was already able in the water to face the land? Had it been without such an ability, moreover, its adventure would have been ill-fated: it would have arrived on land only to be unable to face it. The water creature had an ability to face the land, *first* exercised in the water.

Many loosenings later, the human being arrives. We speak of this arrival as the advent of culture. The root of the term 'culture' is to-till-or-cultivate. The advent of agriculture is a paradigm of the advent of culture. Food no longer just grows in the world; *we grow it*, thereby eliminating any question of searching for a way to face it. Here is the beginning of a profound change, away from the kind of embeddedness in the world that all other things experience. Some ten thousand years later we have loose mouths: our food stores display the same food year round. To distinguish modern life from the life of early human beings, Neumann spoke of the latter as "an indissoluble unity" of "outside and inside" (1974:42). Have outside and inside somehow come to be divided for us today?

The term 'divide' has the root to-force-asunder, and the associated term 'differ' has the root to-carry-or-bear-apart. Here we get the sense of something once whole and now broken into pieces, perhaps its pieces. The term 'separate' has the root to-make-ready-or-prepare-apart. Here we get the sense of something that need not be broken into pieces. It could be one whole with pieces, or rather, with parts that are together as that whole. The making ready or preparing can encompass all that is apart in such a way that the whole of it is thereby made ready or prepared. Hence the difference between the terms 'divide' and 'separate' is this: only the former requires that the relevant whole *be in pieces*, inasmuch as the latter allows that the relevant whole *not be in pieces,* but somehow together as that whole. An order of separation without division is therefore *an order of co-making*, in the sense that the making ready or preparing encompasses all that is apart, a whole with pieces perhaps, but not in pieces, *not broken* (or carried or borne) apart.

In the order of posture a thing and that which stands around it are bound together, at least in the sense that they are conditions for each other: as a thing makes a place in the world, the rest of the world must give place to it, and so forth. The root of the term 'condition' is to-speak-together. Hence a thing and that which stands around it speak together of their posture. Early human beings and that which stood around them spoke together of their posture in an order of separation without division, in a world entirely co-made.

How was this possible? Certainly they had faces that looked very much like ours long before the advent of agriculture, indeed, many thousands of years before the advent of agriculture. How shall we understand a face, especially a human face? A face can be said to bind together the thing that has it with that which stands apart in the world, but at the same time can be said to loosen the thing from the world: are the face and the world somehow loosely bound together, or is the face simply loose, not bound together with the world? This tension marks a face as the very special part of the body that it is.

II. Faces

In his work on ancient skywatching, E. C. Krupp spoke of bees as examples of life forms that orient themselves relative to the sky: bees "use the position of the sun and its polarized light to find their way from hive to flower and from flower to hive." He also associated early human beings with the sky in a significant way:

> The Desana Indians of Columbia even describe the sky as a brain, its two hemispheres divided by the Milky Way. Their brains, they say, are in resonance with the sky. This integrates them into the world and gives them a sense of their role in the cosmos. The perception of the Desana was common to many ancient peoples.(2)

Hence the Desana participated in co-making, in an order of separation without division, through the resonance of sky and brain; sky and brain are apart, or elements of a whole with parts perhaps, but not in pieces, not broken apart. So, too, does the bee participate in such an order, not broken apart from the sun.

In accord with this participation, we find that the root of the term 'attention' is to-stretch-to. Both bees and early human beings paid attention to the sky as if they were stretched to it in an order of separation without division, though the face of the bee and the face of the early human being were still vastly different, a gap that the modern human being has widened even further. We no longer speak of a resonance of us and that which stands around us in the world. (To do so is to refer to a "paranormal" phenomenon.) As we will explore in detail throughout the remaining sections of this part, the early human resonance testifies to our original posture, not the one we have come to assume.

Again, a face works at the boundary inside/outside, stretching to that which stands at a distance in the world. To face is to face out, away from the face, constituting an asymmetry with respect to which all faces are oriented in the world. We can understand this orientation in two, fundamentally different ways: either it is referred to the world, or it is referred to the face. On the former alternative it is all too easy to believe that the face is dominated from outside, as if the world were doing the stretching. But on the latter alternative it is all too easy to believe that the face is dominated from inside and is entirely loose from the world. Here too, for the opposite reason, the face may not retain its distinctive character of stretching *to* the world.

To consider the former alternative, I will make use of a purely behavioristic or, let us say, outside-in analysis of attention: a readiness to respond to stimuli. (Many dictionaries try to *define* 'attention' in this way.) The analysis has always been regarded as most successful when applied to so-called lower life forms, such as B. F. Skinner's pigeons. This success can be explained, though at the same time the analysis will be found wanting.

A thing pays attention to this or that at a distance in the world. Or, on an outside-in analysis, a thing is ready to respond to stimuli from this or that. All the events being considered here are arranged outside of each other in a series, what Skinner called "a causal chain." We have an external stimulus, a thing's readiness to respond to that stimulus, and finally the response itself, an external manifestation of the readiness to respond—hence the notorious S-O-R causal chain. The stimulus must be what it is *outside* the face in question; the face must be ready to respond to the stimulus as the stimulus is *already* constituted. Broken apart from the stimulus, the face must *wait* to be stimulated, hardly in a position to be said to respond freely to the stimulus. It is no wonder that an outside-in analysis appears to be successful for the faces of the so-called lower life forms.

Consider a frog. We now know that a frog has a few basic abilities to use its eyes, one of which is to respond at close range to quite a small object that is moving quickly and erratically. Whenever such an object is near enough, out darts the frog's tongue, and the world is less one flying insect. But if we simply hang many fresh-killed flying insects near a frog, the frog will go hungry (Jastrow, 69–70). The frog, we can say, has an inherent readiness to respond to stimuli from flying insects.

How can the frog afford to have an automatic response of this kind? How can it still survive? Because the frog and the flying insects are not really broken apart, in pieces, but instead participate in an order of separation without division, an indissoluble unity of outside and inside: they co-make an order in which they are arranged apart in such a way that they are still together, making possible the success of an automatic response. A frog cannot stretch to the hanging-dead flying insects because they are broken apart from the frog by a human being, really a modern one, who does so to answer a question about the frog's abilities. Here the break must be taken to be *outside* the frog in such a way that it becomes the very break presupposed by an outside-in analysis: *this* world appears to be in pieces, as if it were a puzzle with a missing piece that could make all the present pieces fall into place. Is a piece missing away from the frog's face?

Before we answer this question let us consider another example of an automatic response. Visible in ultraviolet light to the bee, but not to us, is a special spot on which the bee should feed, "a nectar guide" that evokes the automatic extension of the proboscis:

> One can see that the colors of the flowers have been developed as an adaptation to the color sense of their visitors. It is evident that they are not designed for the human eye. But this should not prevent us from delighting in their beauty. (Frisch, 1976:25)

Nor should it prevent us from understanding this phenomenon as co-adaptation of bee and flower, the former for food and the latter for pollination. (The antenna of the bee is equipped with organs of smell that allow the bee to restrict its visits even further, to a single species of flower, reinforcing both its automatic response and its pollination.) The flower and the bee stretch to each other in such a way that we are almost tempted to say that the flower faces the bee exactly as the bee faces the flower. The orientation of the flower is referred to the bee, and the orientation of the bee is referred to the flower. To understand this mutual reference we cannot allow the world of the bee and the flower to be in pieces.

For bees the most telling use of light from the sun is to orient themselves in general, not just to flowers. They do not depend directly on the sun because direct dependence would demand that the sun be more available than it typically is. On a cloudy day light from the sun is polarized relative to portions of the sky and to the position of the sun in such a way that any patch of blue sky is enough for the bee to orient itself. Bees can perform dances that specify to other bees that food is near by, or that food is far away and in a certain direction. On the latter alternative when bees dance on a horizontal surface in a hive open to the sky through a narrow tube, the dance is oriented only if the tube is not blocked and no cloud covers the exposed portion of the sky. Otherwise the dance becomes disoriented. Hence the orientation of the bees is referred to the light in which they dance.

The light that bees use to orient themselves falls upon all of them at once. The dancing bee does not communicate the direction to fly simply by facing in that direction independently of the light, even though the light itself can be described independently of the bee's face. The bee must face in the light, and when other bees follow suit they are in the same light, thereby learning where to fly:

> The principle whereby direction is indicated by means of dancing on a horizontal surface: during the wagging run the bee takes such a position that she sees the sun [though not necessarily directly] at the same angle as during her previous flight to the feeding place. (Frisch, 1976:112)

An outside-in analysis, on the other hand, declares that a bee's face is ready to respond to a stimulus that is already constituted, as if its face were somehow there, broken apart from the stimulus, waiting to be stimulated. But, first of all, can the bee really be said to wait for the light? Does it have the ability to face without the light? Well, perhaps so, but certainly not with respect to facing its food at a distance in the world. Here the bee is *able to face only in the light*, as if it were not to have a face otherwise: a bee's face is not loose from the light.

And, second, we must remember that a bee stretches to the light, or rather, to the sun through the light; it compasses the sun. The root of the term 'compass' is together-step. The light does not make a bee's orientation; the light is a stimulus only in the world *of* a bee's face. The root of the term 'of' is away-from. The light is a stimulus only in the world away from a bee's face, not simply outside a bee's face, or broken apart from a bee's face: *they together-step*. Although we will not as yet be able to do justice to this change—for that is the task of part two—let us retract our saying above that the light can be described independently of the bee's face. Let us say instead that the light binds together the bee and the flower in such a way that it cannot be understood apart from them; we can no more break away the light from the bee and the flower than we can the bee and the flower from the light.

It is through an intervention in this bond—the root of the term 'intervene' is to-come-between—that a human being answers a question about the ability of a bee to respond to polarized light, and again we must take the resulting break to be *outside* the bee. *This* world appears to be in pieces, as if it were a puzzle with a missing piece that could make all the present pieces fall into place; the light reflected to a bee's face from the particles in a portion of the sky, or from a flower, *makes* the response of the bee, and the bee pays the price of not being free. But the light in the world of the bee's face cannot be removed without, as it were, removing the bee's face as well; only the human face remains to notice anything is missing, let alone that it is a piece to a puzzle of pieces. Just so, a flying insect is a stimulus only in the world of or away from a frog's face, not simply outside a frog's face, or broken apart from a frog's face. In the latter case, again, only a human face remains to notice anything is missing, let alone that it is a piece to a puzzle of pieces. Not to be loose in an order of co-making is not the same as not to be free in an order of making, in a world in pieces.

A human face, early or modern, introduces a kind of loosening from the world, making possible the reference of that face's orientation in the world to that face itself. An early human face is just beginning to be loose in this way. Although the Desana Indians spoke of the sky by reference to their own brains, the sky still resonated with their brains. A modern human face is not only loosened from the world, but also, so it is said, broken apart from the world. Let us look into such a human face.

I am here and I wish to go there. So a friend volunteers to give me directions. First she puts me through some elementary exercises to make sure that I will be able to follow her directions. She says, "About-face!" Then, "Right-face!" and "Left-face!" I pass the test, and so she proceeds to orient my face in the world: she simply faces me there. (Remember playing pin the tail on the donkey?) Then I am to go forward so many paces, right so many paces, left so many paces, and so forth. Not only can I follow these directions, but I can return with them as well. (Or alternatively, if I have no directions but simply go from here to there while keeping track of my paces in the manner of my friend's directions, I can return exactly as I do with her directions.) All along my orientation in the world depends on my original face, and I make use of the world only to sustain that face: first I face the world, and second I refer the world to the orientation of my face. Although I do face *in* the world, I need not pay attention to anything particular there. If a terrible wind spins me around in such a way that my original face is lost, then I too am lost. (Again, remember playing pin the tail on the donkey? Even without a blindfold, an original face is all too easily lost.) But no such wind—well, unless it is awesome—can similarly disorient a bee; when it stops spinning it is still in the light. But when I stop spinning I am not still in the light; I am proceeding only by my friend's directions. To make exclusive use of such directions my face cannot work at the level of the bee's face.

The terms 'left' and 'right' are very special, if designating either of two mirror image sides of some object in the world. Such an object is said to be symmetrical, though it is typically not perfectly symmetrical. Consider the two hands of a normal human body: one is always stronger or fundamentally different from the other in such a way that we can point out which one of our hands is left or right by referring to this difference, a kind of asymmetry. (The *OED* claims that the right hand is usually stronger than the left one.) But here we are considering left and right *hands*, not just the difference left-right, and consequently a certain face—the face of the human being with those hands—is already embedded in them: it is with respect to that face that the hands become left or right. Even if the hands were perfectly symmetrical, each hand itself would be an asymmetrical object, allowing a discrimination between the two. But again, without at least a tacit reference to a face it is not possible to indicate which one is left or right: left or right is of or away from a face in the world, not in the world away from a face, such as left and right hands are.

Dictionaries use ostensive recipes for the terms 'left' and 'right': if I face north, my east side is my right side, and my west side is my left side. This recourse to an asymmetry of the world can be said to resolve an ambiguity that cannot be resolved by the standard tautology, such as that east is the direction of the sunrise. Left and right are directions, but they must be referred to a particular face. This we call "ostensive," and the root of the term 'ostensive'—or of the archaic 'ostend'—is to-stretch-towards-or-in-the-direction-of, or to-stretch-facing-or-in-front-of. Here in the root of 'ostensive' we see again the significance of a reference to a certain face.

To borrow an example from Gregory Bateson (82–3), imagine a man facing north and also shaving with his right hand in front of a mirror. With respect to his mirror image face he is shaving with his left hand, though with respect to both his face and his mirror image face his east and west sides are the same, respectively. East-west and left-right do not work at the same level of orientation in the world. Spinning bees cannot become disoriented because they are referred all along to the level of orientation in the world represented by east-west. Spinning human beings can become disoriented because they are referred all along to the level of orientation in the world represented by left-right. Although the level of left-right does not exclude the level of east-west, it does not require it either, thereby loosening human faces just enough to allow spinning human beings to be lost in the world, in need of recovering their original faces.

Let us think again about following the directions my friend gave me. When I referred the world to my original face, suppose that I chose a certain tree as my landmark. First I faced in the world, and second I referred the tree to the orientation of my face: it was the largest tree, slightly to the right of my line of movement. Does a bee do this with the sun? No, it cannot first face and then refer the sun to the orientation of its face. Or, does the bee face the sun and then turn right or left? No, to put it in our terms, when the man who is shaving turns west, so too does his mirror image: two bees flying toward each other, both turn west. Hence the orientation of a bee's face is immediately referred to its landmark, the sun: they are bound together, or together-step.

Again, what is away from a bee's face in the world is always already in the world away from its face. Here is the key to the bee's automatic orientation to the flower and the sun: what is away from its face in the world is referred to the world in such a way that it is automatically in the world away from its face. A nectar guide is automatically in the world away from a bee's face (though not simply in the world), whereas my tree-landmark is not automatically in the world away from my face. If the tree-landmark were automatically in the world away from my face, I could not first face in the world and next refer the tree to the orientation of my face: as soon as I face in the world, the tree will be my landmark. No, the tree-landmark is away from my face in the world exactly as my right hand is away from my face in the world: the tree-landmark is in the world away from my face as a tree, as the largest tree among others, but *as a landmark it is only away from my face in the world*, slightly *to the right* of my line of movement. (As Sartre knew so well, what must be counted as a stimulus or a cause to a human being may only be away from that being's face in the world, not always already in the world at all.) For the bee the sun-landmark is not only in the world away from its face as the sun but also automatically in the world away from its face as a landmark, for example, slightly *to the east* of its line of movement.

It is worth being very careful here. The additional terms at work in the description of human posture are often found at work on other levels of description. The sense of 'left-right' that can be said to apply to the bee is the one we use in saying, for example, that we fixed the left-front tire on a car. For us this is another example of how we embed our faces in the world. In the common language we use for cars, 'left' and 'right' are frozen so that no one needs to know which way we are facing to understand us. When we say, as we tend to do, that a bee turns left or right, we must be careful to freeze left and right here too so that they are not away from a bee's face in the world but instead away from our faces in the world away from a bee's face.

The same caution must be extended to such terms as 'front' and 'back,' or 'forward' and 'backward.' I have a frontside and a backside, a front and a back, a forehead. The root of the term 'front' is forehead. I face front as necessarily as I face out. When I look in the mirror, my front and back switch, though the front that I have in the mirror is not the front that I have in facing the mirror, as my left and right sides are mirror image sides: again, my right hand in facing the mirror is my left hand in the mirror. My front depends in a subtle way on a particular designation of my left and right sides, with respect to *that* front. Here front takes on the significance of being referred to a face exactly as left-right does.

What is added to my posture, or to the matrix of my incarnation, by adding a front to my already having, say, a nose and eyes? When a cat looks into a mirror it does not see itself at all. A cat cannot switch between its face and the face in the mirror; what is in the mirror is not only not its face, but no face at all. A cat has, say, a nose and eyes but no front in the sense in which a human being has a front. Just so, a bee has a proboscis and eyes at one end of its body but no front. It is our front that is embedded in our hands and cars, becoming frozen there in such a way that we almost forget its origin in our own faces.

Even in its home the bee depends on outside light to orient itself in the world. (As Karl von Frisch pointed out, one species of bee can replace outside light with gravity, utilizing the direction of the pull of gravity as it normally does the direction of the sun: it can face *in* gravity as well. This species of bee, let us say, is more loosely bound together with the world than other species are.) Now I imagine being in my home, and then going somewhere. I get in my car, drive it out the driveway, and turn left or right at the street. From that point I imagine following directions strikingly like the ones my friend gave me above: I go forward so many blocks, right so many blocks, left so many blocks, and so forth. Moreover, the orientation of my driveway may be only an externalization of a certain face of mine, not referred to the world itself, say, by running east-west. Perhaps for me my driveway only runs from the front of my home into the road! My face is already embedded in my home in such a way that, should I be completely cut off from outside, I will still be adequately oriented in the world. (Why not cut off the gravity too? A human being will be oriented, though he or she may well have other troubles.)

We have embedded our faces so pervasively in the world that we often have no idea how the world itself is oriented. When asked we can usually find out, to be sure. But, again, we may very well have no immediate awareness of the answer. We may need to look at the sky: "Ah yes, my driveway runs east-west, and I turn north to go to work in the morning." A bee cannot wait to find out where it is relative to the sun; it does not have a loose face in this respect, a face *to question* the sun. For the bee the difference between map and territory cannot exist, as the bee is not loosened from its territory in such a way that it can need or use a map. We, on the other hand, are so loose that our maps are embedded in the world—constituting the orientation of what Sartre (1976) called "the *practico-inert*"—in such a way that the actual orientation of the world may be superfluous, though the world will always enter into our maps relative to our faces.

It has been said that the first mammals survived the dinosaur age by being able to gather food at night when the dinosaurs, dominated by their eyes, were asleep. These first mammals were dominated by their noses, and to this day "the perception of polarized light is a capability distributed generally among insects and crustaceans . . . [and] also occurs among spiders and even among octopuses and squids . . . [though] among human beings, and among all vertebrates, one looks for it in vain" (Frisch, 1976:131). Hence it has also been said that the first mammals were sufficiently loose from the world to use "a map" relative to odors, orienting themselves at night (Jastrow, 52–4).

Is this right? Could the first mammal make a landmark of a certain odor as we can make a landmark of it? Was the first mammal really loose in the required way? Let us imagine the required way: the mammal first faced in the world, and second referred a familiar odor to the orientation of its face, making an odor-landmark. Shouldn't we say instead that, exactly as the bee must be in the kind of light that orients it, so too must the first mammal have been in the kind of odor that oriented it? If the familiar odor had been eliminated from the world away from the first mammal, wouldn't the mammal have become immediately disoriented? The first mammal faced *in* odor to communicate, and when others of its kind followed suit they were *in* the same odor.

The first mammal's posture was such that it oriented itself in the world with its nose. Its nose was its front in the sense that its body worked so as to place the nose first along its line of movement. Only under severe pressure would the mammal abandon this posture. But this posture does not require the world to be referred to the nose in such a way that the mammal can be said to have a front as we can be said to do. The common argument that the cerebral cortex grew up around the smell brain of the first mammal, that is, that the origin of intelligence lies in the special demand placed on the first mammal by having to rely on odor instead of on light, on its nose instead of on its eyes, may very well be right. (Perhaps the demand is to work with time differentials between the particles arriving at one nostril and those arriving at the other nostril in order to determine the origin of the particles.) But in being right the common argument does not require that the first mammal was broken apart or loose from the world as we are today. Our state is the end result of a long evolutionary process, not its beginning.

We tend to cover our homes and that which stands around them with artificial odor as well as artificial light, and in the process we do not become disoriented, though perhaps we should do so. What is familiar to us in the world may be just our own faces. I get around in the world in such a way, again, that I can return to my starting point by retracing my steps. Some of us are better than others at doing this, and some of us—those we say have a sense of direction—do not even need to keep track of their steps to return as they came. But then again, when it comes to the very unwitting way in which certain human beings know how to return, perhaps they are exercising a different ability altogether, an ability that all of us somehow still share with the bee, the first mammal, and the early human being: an automatic sense of one's place in the world, a bound-togetherness *with* the world. Yet we are more often than not mystified by what sense of direction we do have, and we are truly fascinated when our dogs and cats find their way home from all too far away.

We say there is no place like home, but we can find ourselves in search of our home. For us the rest of the world can take on the significance of a threat or an enemy, potentially undermining the way we refer the world to our faces. Only a life form with such a loose face as ours could have the vaguest sense of this significance. (A human face allows a question just where the bee always already has an answer. It is no wonder that a young human being asks so many questions!) Although the way we make places for ourselves in the world always reflects such loosening from the world, it need not reflect the extreme loosening of being broken away from the world, inasmuch as our original posture speaks of bound-togetherness, should we bother to listen to it. To be loose in an order of co-making is not the same as to be free in an order of making, in a world in pieces.

III. First Human Faces

To contrast the bee with the human being as sharply as possible, I have not yet considered the way in which an individual bee may have a landmark. Frisch carried out experiments that displaced colonies of bees so that "continuous guidelines"—"the edge of woods, a road, a shoreline"—were brought into conflict with the sun. The bees might well follow a continuous guideline, though "individual trees or mere groups of trees were not able to compete with the sun" (1967:342, 346). In *all* cases, the sun still provided the basis for individual bees to rely on a continuous guideline:

> Bees use landmarks [other than the sun] for their *own* orientation. Landmarks are of no account in the orientation of the colony, in the transmission

> of information to the hivemates, because the "language" of bees has no words for them. (1967:333)

First the colony faces the sun in its characteristic way: *each bee has a colony-face.* An individual bee cannot simply face in the world, and then refer a continuous guideline to the orientation of its face; the orientation of its face is already referred to the sun. Hence the continuous guideline itself is first referred to the sun, though indirectly, through a colony-face. Later, after the bee is "trained" to the guideline, the orientation of its face may be referred directly to the guideline, thereby providing the bee with a degree of looseness quite valuable in extreme weather. Such looseness does not require that the bee have a face whose orientation in the world is referred to itself: no bee can have its *own* face. Besides its colony-face, a bee has no other.

Early human beings must have had faces whose orientation in the world could be referred to the world in ways that far surpassed those of the bees. But did they have their *own* faces? Perhaps as the bees, "the colony" first faced the world in its characteristic way: each early human being had only a colony-face. Well, once again to contrast the bee with the human being as sharply as possible, I have not yet considered the early stages of the development of human posture.

From the perspective of current paleoanthropology, our lineage stretches so far into the past that we may well have trouble imagining it. By one and one-half million years ago the lineage boasts *Homo erectus:*

> Put him on the subway and people would probably take a suspicious look at him. Before *Homo erectus* was a really primitive type, *Homo habilis*; put him on a subway and people would probably move to the other end of the car. (Johanson and Edey, 20)

The name of *Homo erectus* aside, our ancestors had had well over two million years of practice at bipedalism, and even *Homo habilis* had *human hands, feet, and teeth. Homo erectus* could have had a cranial capacity nearly as large as the mean cranial capacity today, though by the end of the period of *Homo erectus*, spanning over a million years, "between the habiline stage on the one hand, and the *H. sapiens* stage on the other, the mean cranial capacity of mankind doubled its size" (Tobias, 95). So, too, could we pass on the subway. We have not been "beasts" for a long time, far, far longer than Vico could possibly have imagined.

The lithic technology of *Homo erectus* constituted "a spurt" compared to that of *Homo habilis*, but remained "stubbornly resistant to change for at least a million years" (Johanson and Edey, 229):

> *Homo erectus*, it is fairly clear, evolved practically not at all during that immense time. Then, suddenly, humanity took another spurt. About two hun-

> dred thousand years ago there occurred a second technological leap, and out of it arose *Homo sapiens*. (375)

Shortly thereafter as well, prior to the emergence of "fully modern *Homo sapiens* . . . about fifty thousand years ago" (Leakey, 189), the mean cranial capacity reached its peak, then fell somewhat, before reaching the value it still maintains. At the time of this peak it is also likely that a one-year-old *Homo sapiens* had the cranial capacity of a six-year-old *Homo erectus*, a capacity thought to be necessary, if not sufficient, for "the symbolizing ability" (Tobias, 141). Did we pass through some sort of inflection point between two hundred thousand and fifty thousand years ago?

William Irwin Thompson believed so:

> Marshack claimed that the Paleolithic *baton de commandment* with its engraved markings was a calendar stick for a primitive form of lunar "time factoring,". . . . The implications of Marshack's observations were enormous, for they meant that as early as fifty thousand years ago primitive humanity had observed a basic periodicity of nature and was building up a model of nature. (95)

Thompson went on to argue, against Alexander Marshack, that this "time factoring" must be attributed to women, who were able to correlate their menstrual and pregnancy cycles to the course of the moon. Independently, Theodore Thass-Thienemann has shown that "for the early ages of the Western world . . . the moon-month was the measure of time":

> The immediate reference to the human organism is apparent . . . in this case. Pregnancy was formerly counted not as a nine-month but as a ten-month period. The normal child at birth is called *dashamasya* in Sanskrit, a "ten-month" one. In early Greece the pregnancy year was also ten months. . . . The same was true for the Romans. The ten-month pregnancy year gradually changed, under Babylonian influence, into the twelve-month sun year. (378)

Perhaps, as Thompson speculated, "astrology does not begin (as is often thought) with the Mesopotamians, but goes back to a lunar astrology in the Upper Paleolithic," the foundation for "the lost religion of the Great Mother" (100).

In any event, not until the Upper Paleolithic, about 30,000 B.C. to 10,000 B.C., does the archaeological evidence indicate the final stage of the development of lithic technology:

> They had achieved total control over the process of fracturing, chipping, and shaping crystalline rocks, . . . and they have aptly been called the "master stoneworkers of all times." Their remarkably thin, finely chipped "laural leaf" knives, eleven inches long but only four-tenths of an inch thick, cannot

> be duplicated by modern industrial techniques. With delicate stone awls and incising tools called burins, they created intricately barbed bone and antler harpon points, well-shaped antler throwing boards for spears, and fine bone needles. (Harris, 1977:9)

Besides, they created the first stone-etchings, for example, the *Venus of Laussel*, of about 19,000 B.C., "this Paleolithic Madonna," expressing "the relationship between the moon and woman" (Thompson, 103)—and the first cave paintings, for example, the bird-headed man in the Shaft of the Dead Man at Lascaux, of about 15,000 B.C., "the image of a 17,000-year-old shaman who . . . seeded a dream . . . to make-real the world of the dream" (Honegger, 11). But no matter how they are interpreted, these creations indicate the kind of achievement we expect of "fully modern *Homo sapiens*." Only a relatively short five thousand years or so after the "dream-seeding," we were engaging in agriculture, at the advent of the Neolithic revolution, and in another five thousand years or so, we were writing, at the advent of the Bronze Age.

To use my hand to make my food is one thing; to use something held in my hand to make my food is another; and to use something held in my hand to make another thing to hold in my hand to make my food is yet another. In the middle stage, I use a tool, and in the final stage I use a tool to make a tool. Each stage has degrees as well. Consider one degree of the middle stage, attainable by a chimpanzee: "active transformation [using only natural equipment of *hands, feet, and teeth*] of material having a preliminary neutral form, despite its false outer appearance, into an implement with definite parameters" (Tobias, 122). Here is a degree of looseness: at first the implement with definite parameters is not in the world away from my face. If the orientation of my face in the world were automatically referred to the world, I could not engage in *active* transformation. The material having a preliminary neutral form would be away from my face in the world exactly as it would be away from the faces of others of my own kind; I would have only a colony-face. (Note that a continuous guideline must be away from the face of the individual bee as it would be away from the faces of others of its own kind. No hint of active transformation can be discovered here.) The implement with definite parameters must be away from my face in the world *alone*, referring the material having a preliminary neutral form, if not the world itself, to my face. Perhaps only for some degree of stage three, a stage not at all attainable by a chimpanzee (Tobias, 123), would it be necessary for the material to be away from a face whose orientation in the world could be referred to itself.

Put another way, to engage in the active transformation of a material I must be able to interrogate that material. The root of the term 'interrogate' is to-ask-between. I must be able to make a place relative to that material so that a question arises between us, just where, were I not so loose, *the* answer would always already exist. Any alternative to what is in the world away from my face would be excluded, though what is in the world away from my face might well not be in the world away from the bee's face: each colony faces the material in its characteristic way. Only as *the* answer could the lithic technology of *Homo erectus* remain the same for so long. Perhaps they were even loose enough to transform their lives more actively, but circumstances never warranted it: *the* answer *worked*. (Even today, as Marvin Harris pointed out, a number of societies do not engage in agriculture because collecting and gathering are *less labor intensive*.) The lives of *Homo erectus*, though considerably more sophisticated than the lives of bees, must still have been relatively automatic. Only in the lives of the "master stoneworkers of all times" do we surely find evidence of the fragility of automatic bound-togetherness.

What did it feel like to exchange automatic knowledge for automatic ignorance, to question without *the* answer? To be newborn without forerunners?

> Each of the examples . . . reveals the same "primitive" ontological conception: an object or an act becomes real only insofar as it imitates or repeats an archetype. . . . Men would thus have a tendency to become archetypal or paradigmatic. This tendency may well appear paradoxical, in the sense that the man of traditional culture sees himself as real only to the extent that he ceases to be himself (for a modern observer) and is satisfied with imitating and repeating the gestures of another. (Eliade, 34)

Well, what acts were real? The ones we always already knew to do. We imitated and repeated to recapture a former life, an automatically real life, of making a place in the world for all time: we had always already made our places in the world.

Yet here and there a rupture could occur, leaving an early human being with a question. We most likely did not become loose at once, or loose to the same extent, or loose once for all. During a rupture the gestures of a former life would still be evident; not even a recollection would be required to know what to imitate and repeat. Still, the experience of a rupture could not automatically feel good:

> As for the primitive societies that still live in the paradise of archetypes and for whom time is recorded only biologically without being allowed to become "history"—that is, without its corrosive action being able to exert itself upon consciousness by revealing the irreversibility of events—these primitive societies regenerate themselves periodically through expulsion of "evils" and confession of sins. The need these societies also feel for a periodic regeneration is a proof that they too cannot perpetually maintain their position in what we have just called the paradise of archetypes, and that their

> memory is capable (though doubtless far less intensely than that of modern man) of revealing the irreversibility of events, that is, of recording history. Thus, among these primitive peoples too, the existence of man in the cosmos is regarded as a fall . . . [and] even in the simplest human societies, "historical" memory, that is, recollection of personal events ("sins" in the majority of cases), is intolerable. (Eliade, 74–5)

No member of an archetypal society may face the world in a personal way, *alone; we must not have personal-faces.* Instead, like a bee, each of us must face the world in the characteristic or archetypal way, thereby automatically having a colony-face, perhaps even the very face one would have had before "the fall." *Only through our colony-faces* may we differentiate ourselves, make individual landmarks, and vary our routines.

Let us consider the following analogy, originally employed by Bohm in another way (Bohm *et al*, 1975). First, compare each instrument in an orchestra to a colony-face in an archetypal society: "When the whole orchestra is playing one theme all the instruments are related in an essential way, though each one may be following its own course of notes." A colony-face may follow its own course of landmarks and routines, but only "in a single whole theme" of such courses, perhaps requiring numerous common landmarks and routines, but certainly requiring an order of co-making with the world as well as with themselves. Second, suppose that "the orchestra suddenly begins to break up so that each instrument plays independently (i.e., solo) in a way that is not related to how the others are playing." Or just one instrument, even for a moment, plays solo—a personal-face referring certain aspects of the world, if not the world, to itself *alone* in a way not related to the play of other faces. (The root of the term 'solo' is alone.) Toward the extreme each face *can* play solo—our faces are *first* personal, first personal-faces, our *own* faces, each able to refer the world to itself alone—*loose*—though not necessarily ever attaining the extreme at which each face does play solo, referring the world to itself alone in a way not related to the play of other faces. Even today's society does not, indeed cannot, enjoy such a complete lack of connecting themes, though we do have our own faces.

We reject an archetypal existence, applauding the recollection of personal events in its most elaborate form. Yet it took ever so many years of a patient, archetypal existence before it was possible to "begin to break up the orchestra," and only in the last thousand years or so, especially the last three hundred, did the process finally begin to erode "a single whole theme," leaving us today with our own faces and their associated "individual" landmarks and routines (Aries, 1981). (Did it take all those years to breed a species of animal with the faces to boast: "I never forget a face?" To what extent should we take this question to be about our physiognomy? Consider twins today, but especially this actual case: a black woman and a white man have "fraternal twins where one is black and one is white.") Before the rupture of our archetypal existence we were like the bees, always already related in a single whole theme. But after the rupture—not so sudden or entire as I am again pretending—human beings had *to discover a theme, to find answers.* Vico believed that three institutions—divine providence (or religion), marriage, and burial (1984:§360)—constituted our original answers, based on "a common human sense," not on "one nation following the example of another" (1984:§311). Now we can hardly *ask the questions* at the foundations of these institutions, let alone imagine how it felt to ask them for the first time: we *invented* them!

Doubtless we also must have trouble appreciating the looseness required for recollection itself. A past event cannot be in the world away from my face; were the orientation of my face automatically referred to the world, I could not recollect the event. I would have only a colony-face and the associated archetypal existence, in which I would be said to recollect the event only by default, inasmuch as the world would be away from my face today as it had been yesterday. (A bee does not remember to extend its proboscis on the nectar guide; it simply does so. Imagine a bee who forgets! The automatic extension of the proboscis cannot even be said to be a habit.) A past event must be away from my face in the world *alone*, referring the relevant aspects of the world, if not the world itself, to my face. Hence in recollecting I am as *active* as I am in inventing. Not surprisingly, Vico discovered at the roots of the relevant terms that memory began as one faculty with three aspects: (1) remembering things, (2) altering or imitating things—imagination, and (3) giving things a new turn or putting them into proper arrangement—ingenuity or invention. I "make" recollections by remembering, images by imagining, and new turns or proper arrangements by inventing (Vico, 1982: 67–70; 1983: 127; 1984: §819). Or again, more generally, in referring things to my face *alone*, I interrogate those things, and among the possible answers are recollections, images, and new turns or proper arrangements. The most profound indication of looseness is to "make" questions.

But this discussion still misses the mark in a crucial way. The problem is to understand how an *early* human face was in the world. Vico's warning even he had trouble heeding:

> Now, since the human mind at the time we are considering had not been refined by any act of writing or spiritualized by any practice of counting or reckoning, and had not developed its powers of abstraction by the many abstract terms in which languages now abound . . . we can now scarcely understand and cannot at all imagine how the first men thought. . . . For their

> minds were so limited to particulars that they regarded every change of facial expression as a new face, as we observed above in the fable of Proteus, and for every new passion they imagined a new heart, a new breast, a new spirit [or rather, a new head]. (1984: §699–700)

The early human face was loose, its orientation not automatically referred to the world, allowing "a power of abstraction" that was simply "not developed." How "not developed?" We know that later "there appeared in the market place as many masks as there were persons (for *persona* properly means simply a mask)" (1984: §1033). Earlier there must have been a new mask, a new *persona*, for every new face, and vice versa.

> To wear a mask, therefore, is something very different from a game. It is among the most serious and weighty acts in the world: a direct and immediate contact, and even an intimate participation, with the beings of the invisible world, from whom one expects vital favours. The individuality of the actors gives place momentarily to that of the "spirit" which he represents, or rather they are fused together. (Levy-Bruhl, quoted in Cazeneuve, 84)

Originally at every turn, nowadays at some—an extraordinary Halloween, perhaps—if what is in the world away from a human face is new, so too may that face be new.

Early human beings must have participated in a co-making—not just a "making"—of things and faces. Neumann again added just the right perspective:

> Early man lived in the middle of this psychophysical space in which . . . world and man . . . are bound together in an indissoluble unity. . . . Early man spoke no more of *the* sun than he did of *the* moon. Just as he knew a new moon, a full moon, a waning, and a dark moon, he looked upon the eastern sun of the morning, the meridian sun of the noon, and the western sun of the evening as different suns. (1974: 42, 56)

Here was looseness, "a power of abstraction," in the service of "particulars" to a new degree that "we cannot at all imagine." The key must be that, if the "world" was new, so too might the "man" be new. The sun and the face still took together-steps, but now the face was a *human* face: the together-step was new-sun/new-face, or rather, new-sun/new-personal-face. At first our answers to the questions that eliminated automatic bound-togetherness were, just like the question, *between* things and faces. Our active transformations were necessarily as much of ourselves as of the world, for though bound-togetherness was no longer automatic, it was still in sway; we had experience with evident co-making. (Recall how the Desana spoke about a resonance between the sky and their brains.) Only later did we reach experience in which the "power of abstraction" was "developed," and we *retained* personal-faces, *first* personal-faces, our *own* faces, whose orientation in the world could be referred to themselves.

Of course, in the world away from an early human face was only *the* sun. The bee who dances in the polarized light automatically knows *the* sun. No question requires the bee *to learn* of *the* sun: the bee already knows *the* answer. But early human beings did not know. The questions they "made" required them *to learn* of *the* sun, and they etched their answers in stone, doubtless away from the same personal-faces they put in the light of *the* sun. Never more than one sun was away from such a face, because the sun away from that face and that face were a together-step. These together-steps constituted the ground for a recurrence of *the suns* (and the faces), for no face was away from each one in turn. More than one sun/face existed long ago.

So too did more than one moon/face, which both Thompson and Thass-Thienemann declared to be original. Again together-steps constituted the ground for a recurrence of *the moons* (and the faces), for no face was away from each one in turn. The original "lunar astrology" could not have involved what we would think of as "counting the moons" (Marshack, quoted by Thompson, 96) or as quantifying the course of *the* moon. We can see the residue of this lack in our old ways of speaking about the seasons:

> There were all kinds of seasons. In Old English, for instance, there was a *sol-monadh*, "sun-month," about February; a *hyld-monadh,* "stormy-month," about March; a *tri-milch-monadh,* "three-milk-month," about May when the cows were milked three times a day. . . . These terms show how the year, divided into "moons," was not quantified, but perceived in the concrete reality of weather conditions and seasonal human activities. . . . The word *season* developed its meaning from this concept of the "appointed" proper time for action. Seasons return infinitely, as does everything in the "wheel of time." (Thass-Thienemann, 380)

But more specifically, as to the "concrete reality" of women in the Upper Paleolithic:

> Recent studies in physiology have noted that women who live in close proximity to one another—nurses in a hospital or coeds in a college dormitory—tend to have their menstrual periods at the same time. Other studies have shown that women living near the equator have a marked tendency to ovulate during the full moon. It is reasonable for us to expect that women living together in small hunting and gathering bands would all have their menstrual periods at the same time. Since classical myths have associated the menstrual period with the darkness of the new moon, it seems also reason-

> able to assume that the women would be having their periods in synchrony with the phases of the moon. (Thompson, 96)

Hence, the Maoris speak of menstruation as "moon sickness," and of the moon as "the true husband of all women, because women menstruate when the moon appears" (96–7). But, again, Thompson's conclusion that "women were the first observers of the basic periodicity of nature" is itself much too modern to be justified; right as it may well be about the role of women, it presupposes that they *retained* personal-faces, putting them in each moon in turn. Instead, their personal-faces were together-stepping with the moon in order of co-making, made possible by a looseness not experienced until then, a looseness to question *the* moon without yet *the* answer.

When we look back on the lives of early human beings, we tend to think as Vico did: they were "dolts." Were they really dolts? Harris said "no," referring us to their various accomplishments, especially their technology. (Like Richard Leakey (158), he quotes Marshall Sahlins' remark that early human beings constituted "the original affluent society.") But this answer does not meet the question at its foundation. Compared to other beings with faces, including modern human beings, early human beings should be called "the master faceworkers of all times." *They invented practices for achieving certain faces*, at first their original face but later, as we shall see more than once in what follows, the personal-faces required for other *active* transformations, such as reading by speaking. Their practices worked, though they always tended to retain personal-faces, thereby extending powers of recollection and establishing the irreversibility of events. We are too attached to our own faces to imagine what it would be like to engage in such practices, especially to practice *as a community*, aiming together, or aiming to be together, in a single whole theme.

IV. First Promises

Consider this report by Bateson:

> You deal with broken promises in Hawaii with ritual precautions. Something can rub off on you, because every promise contains a curse. You can't get a Hawaiian to promise to come and do your gardening work on a Saturday for this reason, and in old Hawaii they did not make promises. (Quoted in Laing, 1971b: 37)

This, too, must seem paradoxical to us, for now promises are made as routinely as in old Hawaii they were avoided. What is it about promises that made them so problematic?

If I am automatically real, I always already know what to do. But if I am loose in some respect, I am no longer automatically real in that respect. Just at the moment that I always already knew what to do, I *can* "make" a question; I *can* remember *or* forget to act. One way to ward off forgetfulness is to lead an archetypal life in which the past and the future are identical. Hence the intolerable recollection of a personal event: such an event breaks the cycle, forcing a future unlike the past. What *needs* to be recollected does not recur. Or again, a promise is a curse because it assumes that the future need not be like the past; I do not *need* to promise, for I will do what I have always done, indeed, what *we* have always done. If I promise I must intend to sin, to act in a way that I must recollect as a *personal* event.

But given the possibility of remembering or forgetting at almost every turn, I *must forget* at some of these turns as well. If I *always remember*, as Friedrich Nietzsche so aptly put it (58), I cannot "have done" with anything, "so that it will be immediately obvious why there could be no *present* without forgetfulness." The archetypal life must have been designed to eliminate the "*present*": we could not promise the future because we could not "have done" with the past, the paradise of archetypes. Nowadays this must seem terribly remote, for we *only* recollect the past, which we believe to be growing ever more different from the "*present*," even though, as Fernand Braudel concluded (316), the notion of "keeping up with the times" did not exist until about three hundred years ago! (Braudel quoted Paul Valery: "Napoleon moved no faster than Caesar" (429).) We have forgotten the origin of our ability to promise the future, the origin of our will.

Here Nietzsche would have us recall what must go into a simple "I will":

> To ordain the future in advance in this way, man must first have learned to distinguish necessary events from chance ones, to think causally, to see and anticipate distant eventualities as if they belonged to the present, to decide with certainty what is the goal and what the means to it, and in general be able to calculate and compute. Man himself must first of all have become *calculable, regular, necessary,* even in his own image of himself, if he is able to stand security for *his own future,* which is what one who promises does! (58)

Well, it is one thing to be *calculable, regular, necessary* by default, through a colony-face, and another thing to be so by engineering in which we "make" our "own images" of ourselves, through personal-faces. Actually to promise, again, presupposes just the looseness that was characterized so uniformly at our origin as "the fall." Before "the fall" we were *calculable, regular, necessary* by default. Between "the fall" and the "I will" we must have come to retain personal-faces, our *own* faces, thereby providing the foundation for the engineering that Nietzsche alluded to above. Hence our "prehistory" was at an end, leaving us at the dawn of "history," in the posture of making a place one day at a time, as we say. Today I promise I will make the same place tomorrow. The dawn of "history" was the beginning of the series of promises that founded both the irreversibility and the regularity of events.

We need to look at the old way of handling a broken promise. What sort of relationships did we have before we became *calculable, regular, necessary*? Here is Bateson on the practice in old Hawaii:

> In old Hawaii the correct thing to do . . . is to have a "Ho'o Pono Pono." This is a gathering of the entire family, which may comprise several households of married siblings and offspring. In this meeting each member is asked to voice everything he has against every other member of the group. Having voiced all the complaints he can think of against members of the group, he is asked by the meeting's chairman . . . :"Do you disentangle him?" To which he must reply: "Yes." (Laing, 1971b: 37–8)

Obviously each complaint had also been a secret from at least some of the others; hence the point of *sharing* complaints in front of the *whole* group involved in the life of an unfortunate person. For a person to have misfortunes stemming from a broken promise, the whole group involved in his or her life must be tangled, with the broken promise itself being the base of the tangles. We untangle ourselves by eliminating our secrets, especially those constituting a way we are against each other. The goal of a "Ho'o Pono Pono" is *the posture of friendship*, making a place in the world *in the open*, as we say.

What does friendship have to do with promises? Vico provided the clue we need here: "for lack of reflection," our human ancestor "does not know how to feign (whence it is naturally truthful, open, faithful, generous, and magnanimous)" (1984: §817). Were the orientation of my face in the world automatically referred to the world, I would be in the open, allowing another human being to put his or her face in me. Feigning requires a certain looseness: "what I really am" cannot be in the world away from my face, but must somehow be away from my face in the world *alone*. *A secret is a paradigm of looseness*: I "make" it. Or rather—and this must be the key to a "Ho'o Pono Pono"—we "make" it. Originally a secret must have been regarded as *a tangle, between us*, necessarily involving the *whole* group in an order of co-making. A "Ho'o Pono Pono" was addressed to looseness itself, aiming to achieve the friendship of being in the open that looseness always threatened, the friendship we formerly had by default. Now we could not help but get tangled, whereas formerly we were always already untangled. Promising came into its own when it was no longer possible simply to engage in the friendship of being in the open. When life is *first personal*, promising establishes the bonds, the connecting themes, that were originally automatic, embedded in a single whole theme; the "I will" can be understood by another of my own kind, *a shared secret*. (The "we will" can play an equally important role here, our way of establishing as much of a single whole theme as we can do when we are first personal. Notice again that a colony-face always already involves such a theme.) For us "honesty is the best policy," but a policy is grounded in a promise: "I will be honest." Originally we were always already honest, automatically making our places in the open.

For Vico "the true natural friendship" was marriage, which he dated after 3,000 B. C. Here friendship must have its modern form: at the base of the institution of marriage is a promise. Vico claimed that only with "the first thunderbolts after the flood" did "the first men" give up "the infamous promiscuity [*comunione infame*] of things and women" (1984: §504, 553). (Here Vico was looking at human history, as he always did, through language, really, written language. He had no archaeological evidence, though it has been said that such evidence exists for the date of the flood, around 4,000 B. C. But most of all Vico had no idea of the thousands and thousands of years that came before his story began. He could not have grasped the significance of our "making" a question to which *the* answer is biological paternity. Once we always already knew *the* answer, then we did not, and then we did again; we *learned* it, perhaps, as Thompson believed, in the late Upper Paleolithic, but not before. Was this feat in the background of Vico's "mythos?" Recall that between conception and birth we make our places in the world *inside our mothers*, allowing biological maternity the more evident status. Between always already knowing *the* answer and *learning* it, a man would be regarded as necessary to conception, but "not as a cause of it," which cause might well be "the moon" (quoted by Thompson, 125–6). Our original answers upheld an order of co-making extending well beyond man and woman, an order we have forgotten, or alternatively, as Thompson believed, an order *men* do not wish us to remember. Was Vico such a man?) After "the infamous promiscuity," we performed "acts of human love under cover, in hiding" (Vico, 1984: §504). Well, certainly such a way of making a place in the world is not in the open, though it is loose enough for the institution of a marriage promise, a promise that "contains a curse" as well. The first human lovers were *put in the wrong* by the first thunderbolts, leading them to go "out of the sight of heaven," "in shame" (1984: §504). Why not say instead that these "acts of human love" were performed—the first to be so performed—as *personal* events, ones to remember, as we say? Hence, in an archetypal society they must be "sins." To this day "acts of human love" are performed in private, not in public, not in the open; we hide our personal love, and even hidden it is often still "a sin."

If we choose not to follow Vico in regarding "man as divided from man by heart," what path shall we choose? Did Vico use his own evidence? Can the heart that divides be the heart that is new with each new passion? Now we retain our *own* hearts, and can no longer imagine that once a new passion required *two* new hearts, one for each party: hence *passion was co-made*. We believe that our passion need not be returned; we believe in love triangles, presupposing that one party can have two passions *with the same heart*, a heart over which the other two parties may well quarrel, to win it, as we say. Vico also believed in "quarrels produced by use in common" (1984: §553). But such quarrels presuppose a certain looseness; what is always already held in common cannot provide a ground for quarrels. First we must raise *a question of use*, and then we can quarrel about the answer.

What was a quarrel without the ability to feign? The ancient Romans regarded aliens as "eternal enemies at war" (Vico, 1984: §638). They were always already enemies; they were enemies in the world away from their faces. Aliens were also thought of as "robbers," apparently a status of no small worth, "a great honor." Originally quarrels were eternal; a new quarrel must have required *two* new hearts, each set against the other rather than for the other, as in a new passion. And yes, we would have engaged in profound violence, even cruelty, often depicted all too vividly in the old epics. Still, unlike today, the cruelty could not have been corrupted by feigning an innocence that scapegoated the opponent: now it is said that our enemies make it necessary for us to defeat them, bringing the war down upon themselves. Our wars are not co-made, with mutual honor.

We can even recognize such mutual honor in Harris' favorite examples of violent traditional societies today:

> (1) As soon as someone gets killed, there is a truce. For a day or two, all the warriors stay home in order to carry out funerary rituals or praise their ancestors. But if both sides remain evenly matched, they soon return to the fight ground. (1974: 64)
>
> (2) Referring to her own experience, Dr. Shapiro says that her unscarred and unbruised condition was a source of concern to the Yanomamo women. She says that they decided "that the man I had associated with did not really care for me enough." (1974:91)

Harris placed these practices in the context of "reproductive pressure," imagining thereby that he might account for them as aspects of a pattern of behavior that maintained a population balance. Here Harris confused an order of co-making with a world in pieces that he needed to put together. Without having to defend the particular order of co-making, we may still understand that violence can *establish* respect and honor if it is *between us, co-made*. (Commenting on (2) Harris claimed: "While we cannot conclude that Yanomamo women want to be beaten, we can say that they *expect* to be beaten" (1974:

91). But on this conclusion (2) makes no sense. If they do not want to be beaten, then they also do not want "care for" themselves.) It was the same kind of respect and honor that American Indians accorded to the buffalo by wasting no part of their kill (Boyd, 9, 62, 184). Only when killing is not embedded in an order of co-making can it turn into a practice the American Indians never understood: killing buffalo "for sport."

Social historians such as Foucault and Sennett have found that, even after the times we are considering, though no longer, sexuality was a matter of relationship first: originally no one started out with one's own heart, one's own sexuality, and then entered intersexuality. Rather, the opposite was true, exactly as in the case of enemies at war. When we fall in love today we still feel as if we had always already known each other; we do not really believe it. To be eternal lovers, our places in the world must be made together, *without question*. But now it is ever so hard to resist a question of our love, which in turn requires a promise, a marriage. Originally lovers had a future identical to their past. We can still see the residue of such archetypal eternity in the roots of our own language:

> The Old English *aew* . . . means "law" in the sense that we speak of "natural laws." Law, in its original meaning, was not some human statute, but a timeless order, the "divine law" implied in all existing things. . . . The Old English *aew*, *ae* also meant "wedlock, marriage," "spouse, wife." Marriage and wedlock are thus conceived within the framework of divine or natural laws, of the periodic change of generations, therefore, marriage of man and wife represents "eternity" here on earth. (Thass-Thienemann, 384)

Are we able to grasp what this meant? Can we imagine the ancient origin of our "romantic" notion that lovers are "meant for each other?" It is ever so hard to do when one can love *alone*.

It is now thought that relatively stable pairing is very old, perhaps one or even two million years old, though stable pairing should not be confused with monogamy or marriage, of course (Gough; Leakey). No convincing evidence has been found for original group marriage, nor for original patriarchy or matriarchy. The favored, but still moot, conclusion is that matrilinearity was original: typically a mother and her brother shared a certain prominence, not to be thought of as authority or power (Thompson; Harris). As Eleanor Leacock put it, the data reveal "the dispersed nature of decision-making in pre-class [or egalitarian] societies" (20). Intrasexual and intersexual friendships were equally original, the warp and woof of original community. In this regard Leakey concluded:

> When our ancestors took up organized hunting and gathering as a career [at least a million years ago], therefore, one of the major changes in their way of

> life, and one which was to open up a vast behavioral gulf between humans and our closest relatives, was the adoption of [*active*] sharing. This new and unusual form of primate behavior was one of a whole group of traits acquired through hunting and gathering that helped to push our human ancestors towards an increasingly adaptable life. (140)

And at the other extreme was Harris, speaking of "the search for personal privacy" in traditional societies. It is no small task for a modern human being to grasp how it feels to make a place in the open, with "no closed doors," when it comes to matters sexual, to "irritating gossip" (1977: 12). We *invented* "closed doors." (I recall an Eskimo who told me that on moving to a modern city she had to become accustomed to visitors knocking on her front door rather than merely entering her home.) When one is *first personal*, a traditional society is "suffocating" (Aries, 1962). Whereas when one is *first in the open*, hiding may be difficult, but it must be an *adventure*, fraught with danger or risk and excitement. Perhaps we can just barely recognize in our offspring today what it meant originally to develop the ability not to be in the open, for as adults we tend to regard sexual hiding as "cheating." In this light, pairing may have been more stable long ago.

The only intrasexual relationships that Vico (1984) celebrated were between "the fathers," the "pious-strong," who were able to take women and remain "hidden and settled in their fields." They were responsible for the first taming of the land, the cultivation of the fields, burned out of the forests. The first "booty" was the produce of such fields, taken by the "impious-nomadic-weak," whom the fathers defeated, incorporating the survivors into their "families"—the origin of the term 'family.' The "first authority" was the fathers' actual possession of their land and the members of their families, including their wives and offspring. But prior to the first authority, whether or not Vico correctly identified its bearers, the land was always already held in common in an inclusive dominion of "dispersed decision-making," *without greed*, as Karl Marx (1967) put it. An "indissoluble unity of world and man" could not support an exclusive dominion of the land. First, one must be loose; one must have a personal-face to put in the land, a practice a father continued by being buried in *his* land, sometimes along with *his* family. (Burial must reflect some sense of the irreversibility of events: we *invented* it too, perhaps in answer to a question as to how to stay in existence. In his interpretation of Jesus as one magician among others, Morton Smith referred to a magical practice to *achieve* immortality, which we obviously engaged in before immortality became what we believed we always already had. Even closer to our day, in the Middle Ages, we believed that the Judgment Day was "the last day of the world, at the end of time," displaying our "deep-rooted refusal to link the end of physical being with physical decay" (Aries, 1981: 33). Were the fathers—or whoever were the first to be buried—the first to stay in existence through an appropriate burial, thereby also being able to continue possession of the land in which they were buried?) If the orientation of my face in the world is automatically referred to the world, that is, if I have only a colony-face, I cannot possess anything: *we* already possess it. (To this day dictionaries include an inclusive sense of possessives.) To be my *own* possession, a thing must be referred to my face in the world *alone*.

The fathers must have been in a position, then, to promise protection to the members of their families; the first "asylum" was a family, a family-state. No doubt this practice must have originally been almost entirely "mute," without the "abstract" terms that abound in our promises of protection. (The "first men" worked in terms of "a mute language," as Vico put it (1984: §32), "of signs and physical objects having natural relations to the ideas they wished to express.") Eventually the promise of protection became less "mute," however, as the fathers came to be pressured by their growing families to unite in "a theocratic aristocracy." Their "auspices" were still held in secret, expressed in "a secret language as a sacred thing" (1984: §32); that is, the auspices were sacred and "the plebians" were excluded from them, especially from a marriage under them, a proper marriage. The new asylum was a city, a city-state, but the grounding promise remained essentially the same. Again the connection between secrets and a promise: far from untangling the members of the families, the new asylum established another way they would remain tangled, a certain regularity of tangles. The plebians in turn eventually contested the fathers' authority by claiming that the fathers "were not descended from heaven; that is, that Jove was equal to all" (1984: §415). But, as Vico himself noted, the result was only another regularity of tangles: "in the free commonwealths all look out for their own private interests" (1984: §1008). Hence, we did not return to the simple friendship of making places in the open, for the open would become the public, and first we would make places in private, behind "closed doors."

V. First Human Voices

A dancing bee wags her tail while facing in the light of the sun. A number of bees follow her closely within antenna range. In this way the first bee communicates to the other bees, using "a word" in their "language" (Frisch). Notice that "the listeners" come to understand "the speaker" merely by imitating her: they face as she does in the light of the sun. Let us say, "the speech" is ***naturally transparent.*** Imitation alone suffices to understand it, as it is immediately referred to the world. All living things who communicate to each other can do so in this way, a characteristic aspect of the way they make places in the world: their posture and that which stands around them *speak-together.* Perhaps only modern human beings can *speak alone*, loose from that which stands around them.

For Vico, the fathers "were sages who understood the language of the gods expressed in the auspices of Jove" (1984: §381). The root of the term 'auspice' is to-see-the-birds, and in the *Iliad*, after the age of gods, Homer still celebrated the wisdom of those who "scanned the flight of birds" (13). Vico took Plato to task for believing that our original vocal imitations were more a matter of esoteric than poetic wisdom, but Vico still supposed that we imitated certain aspects of the auspices of Jove, especially the sound of Jove: "Thus it is not beyond likelihood that, when wonder had been awakened in men by the first thunderbolts, these interjections of Jove should give birth to one produced by the human voice" (1984: §448). (The American Indian Black Elk also spoke eloquently of the times he heard "thunder-beings" (Neihardt).) Even after Homer the ancient Greeks were fascinated with *mimesis*:

> It meant much more than mere "imitation." In fact, its meanings in Greek range from "copying, fascimile making" to "re-presentation" and even "re-embodiment". . . . The ancient Greeks, at any rate, were clearly addicted to [sound-*mimesis*]. (Stanford, 99–100)

They believed indeed that *sounds* were "the best medium for *mimesis*!" Aristotle regarded the voice as "the most mimetic of human faculties," and W. B. Stanford urged his modern "eye-readers" to have "a greater degree of tolerance for [ancient] ear-readers" (101, 116). Obviously we no longer have the Greek addiction to sound-*mimesis*, nor can we easily imagine their experience. (We say, "A picture is worth a thousand words.") Originally human voices must have been naturally transparent, allowing no division between an act of imitation of a thing and what we would regard as the real thing. Hence, the act was *sacred*: *the* sound of Jove. With this beginning, the later Greek fascination with *mimesis* is understandable.

If the original aim of speech was the real thing, we should think of it more as divination than as imitation; hence, the sense of *mimesis* Stanford called "re-presentation" and "re-embodiment." Certainly Vico celebrated this aspect of the fathers' speech (1984: §381). They would have appeared to us to act as if they were in touch with the real thing, even when the real thing was "invisible" to us. Hence, our inability to imagine, let alone to accept, their experience:

> Primitive men . . . feel themselves *in immediate and constant touch with an invisible world* which is no less real than the other [visible one, limited to nature]; with their dead, recent or not, with "spirits," with more or less

> clearly personified powers [of things in nature], and finally with the many types of beings which people their myths. (Levy-Bruhl, quoted in Cazeneuve, 67, my italics)

Well, "an invisible world" to Lucien Levy-Bruhl at any rate, who doubtless was trying to refer it to his *own* face. For "primitive men" it was as if nothing were really able to go out of existence (even without the appropriate magical practice). Was there any such thing as natural opacity? First natural transparency had to be limited, a rupture of the life in which we spoke-together with the world.

Obviously "the secret language" of the "nobles-priests" of a theocratic aristocracy was not a "mute language," always already in the world away from their faces. Nor, however, was it "a vulgar language," "introduced by the vulgar, who were the plebs of the heroic peoples"; a language the Latins "properly called vernacular," "suitable for expressing the needs of common everyday life in communication from a distance" (Vico, 1984: §439, 443). The latter language must have involved "a common script," and "the vulgar letters had not yet been invented by the time of Homer," that is, in "the age of heroes," when the fathers banded together to form the second asylums, the cities (§429). Vico's threefold division of languages followed that of the Egyptians, "handed down to us as the three periods through which the world had passed up to their time," the third period, "the age of men" (§31):

> The second kind of speech [after that by mute acts], corresponding to the age of the heroes, was said by the Egyptians to have been spoken by symbols. To these may be reduced the heroic emblems, which must have been the mute comparisons which Homer calls *semata* (the signs in which the heroes wrote). (§438)

Hence between the mute language of "hieroglyphs" and "the human language" of vulgar letters came the secret language of *semata*. Jean-Francois Champollion, the first to understand second-age Egyptian hieroglyphs, called them "phonetic symbols," "without being strictly alphabetical yet phonetic" (Ceram, 122).

We use expressions like "what the speaker said went in one ear and out the other." Notice how the sounds must come and go here, as if they were being pulled on a string through the listener's head; the listener puts his or her face in each one in turn, only to forget them in turn as well. Here irreversible time and forgetfulness are essential. Promising is now at work in our speech in a way that it could not have been originally, indeed, *need not* have been originally. Natural transparency took the place of the promising that gradually allowed speech to become more articulate or "vulgar," grounded in our *agreement* to speak in a certain way. (Notice how repetition undermines the transparency of today's speech.) Hence, a certain feigning became possible, the imitation of one's *own* voice: we *tell* lies now! It is no wonder that we fail to understand original *mimesis*!

How did our voices sound before they became "vulgar?" Vico concluded that "the first languages were formed by singing" (1984: §230). The "formulae in which the laws were expressed in ancient jurisprudence were called *carmina*, or songs" (§1036). Even into the times of Cicero the "children learned the Law of the Twelve Tables by singing it *tanquam necessarium carmen*, 'as a required song' " (§469). And still today "mutes utter formless sounds by singing, and stammerers by singing teach their tongues to pronounce" (§228). Vico saw similar behavior in "men moved by great passion" and in young children. Speaking a language "almost entirely articulate and only very slightly mute, there being no vulgar language so copious that there are not more things than it has words for" (§446), is a skill we take so much for granted that we cannot imagine what it was like to acquire it without forerunners, to *invent* it. As Vico so aptly put it, "they had not yet heard a human voice" (§454).

Two hundred fifty years after Vico, Stanford reached a similar conclusion:

> "The science of public oratory is, after all, a kind of musical science": so remarks one of our best helpers, Dionysios of Halicarnassos. On the same principle Quintilian strongly emphasized the necessity for orators to be familiar with both the theory and the practice of music. Like a singer, the speaker should be a master of melody as well as rhythm if he wants to win his cases. To illustrate the importance of a skillfully modulated voice Quintilian recalls that a celebrated Roman orator when making a speech used to keep a man with a pitch pipe beside him to warn him by sounding it if his voice went off the best pitch-levels. (27)

These remarks concern the age of men, not the age of heroes in which song must have pervaded the voice even more. As Vico himself might have put it, an eternal property of singing is that the voice is brought to rest on each note in a way almost entirely lacking in vulgar speech, which is a much more continuous movement (Stanford, 28). Given the singing movement of the pitch of the voice, we separate the sounds—without dividing them—in a way that makes them more accessible in both speaking and listening, that is, more resolved in the sound alone, in the world away from our faces. Hence the sound of ancient Greek:

> As Aristoxenos [a pupil of Aristotle] saw, there is a kind of song-melody in speech, especially in ancient Greek with its melodic word-accent. A later musicologist, Aristides Quintilianus, developing this notion, spoke of a tone

> of voice lying between singing and speaking. He described it as "the intermediate voice-movement, the one in which we make our readings of poetry"—an important indication for us of how we should recite passages from ancient poets . . . a way of speaking which pronounced the Greek pitch-accents more like musical notes than ordinary speech-tones. (28)

Stanford's *mimesis* of ancient Greek speech immediately suggested to me other speech that had resisted corruption by "vulgar" tones, especially the speech of certain American Indians. In ancient Greece, too, the "word-melody" quality of living speech had begun to fade, when, in about 200 B.C., Aristophanes of Byzantium at Alexandria invented the pitch-accents to preserve the melodies, which Dionysios put even before rhythms as "the sources of beauty and charm in composition" (32–3).

"Certainly the founders of Greek humanity," according to Vico, "were theological poets ['poets' is Greek for 'creators'], and these were heroes and sang in heroic verse"; "among all nations speech in verse preceded speech in prose" (1984: §235, 468). To this day the quality of song and verse penetrates us deeply. We still can remember something better if we put it into a melodic verse, in which both melody and rhythm carry us from sound to sound, unfolding the sounds from a single whole theme, separate but not divided. Without our modern power of recollection, the heroes "preserved their language by the oral tradition of their poems," and "the rhapsodes, who, being blind, whence each of them was called *homeros*, had exceptionally retentive memories, and, being poor, sustained life by singing the poems of Homer throughout the cities of Greece" (1984: §466, 878).

The practice of the *homeros* assisted the transition from mute to articulate language and speech. We can easily imagine the listeners trying to sing along as we still do when we hear a song. Vico did not believe that our ancestors became loose at once, or loose to the same extent, or loose once for all:

> For in the state of the families [in the age of gods], which was extremely poor in language, the fathers alone must have spoken and given commands to their children and *famuli*, who, under the terrors of patriarchal rule, . . . must have executed the commands in silence and with blind obsequiousness. (1984: §453)

The terror aside, why not say instead that only those who had a sufficiently loose face, man or woman, could *speak*? Even naturally transparent speech requires a certain *active* transformation of one's body, unless, of course, one has only a colony-face. Vico must have been thinking of the advent of certain active transformation, for otherwise a colony-face would have sufficed, and every member of a family could have spoken long before the age of gods. (With a colony-face we spoke with *one voice*, as we still say. Even into relatively modern times we sang-together, the same part, without harmonies.) So, *whoever* spoke had a personal-face, thereby earning authority that Vico attributed to the fathers alone.

Even today, as Julian Jaynes recognized, a human voice earns authority:

> Consider what it is to listen and understand someone speaking to us. In a certain sense we have to become the other person; or rather, we let him become part of us for a brief second. We suspend our own identities, after which we come back to ourselves and accept or reject what he has said. . . . To hear is actually a kind of obedience. Indeed, both words came from the same root and therefore were probably the same word originally. This is true in Greek, Latin, Hebrew, French, German, Russian, as well as in English, where 'obey' comes from the Latin *obedire*, which is a composite of *ob* + *audire*, to hear facing someone. (97)

And if for us to hear a human voice is actually a kind of obedience, imagine what it would have been like for someone who had no way to "come back to oneself," that is, who had no personal-face to which to refer the voice. Should I as yet not have a personal-face of my own, or rather, *on my own*, I may gain one from another who does have one if I understand his or her speech. Again as Vico might have put it, an eternal property of speech is that a mute may understand it, even when it is not naturally transparent. (Otherwise, of course, we could not *learn to speak*, though not all mutes who understand speech learn to speak themselves.) Here is the true origin of the natural authority of a human voice.

But speaking was also a change of facial expression. So, on a deeper level, could a new face come along with every new speech? Why not think of speaking as significant in the same way for the speaker as for the listener, especially in the age of gods: *two* new faces were involved in every new speech? Hereby the repetition of speech-songs would be indispensable, as one way to practice having the same face, a personal-face, again and again—a stage in the development of the ability to retain a personal-face. We can imagine the secret sessions of the fathers in which they all aimed to achieve the same face by singing together. Or can we? A shared personal-face? As if each singer were to wear the same mask, perhaps? Even today chanting weds the chanters in two quite different ways. For the first way, the chanters lose all sense of the difference between their voices, as if *one voice, one chant*, were away from *one face* in the world. But for the second way, the chanters are *immediately* referred to the chant, undermining personal-faces. (Without the second way, chanting could not have been practiced by anyone who did not already have a personal-face; that is, it could not have been original.) On one side of chanting we have personal-faces; on the other side we do not.

If once every new speech came along with a new face, it is no wonder that the practice of reading with the mouth closed was "rare" until the Middle Ages (Stanford, 1). The families had no writing to read, and even thereafter, at the peak of Greek achievement, the usual practice was to read aloud. If by then we were not loose from our voices, no one in the families could have been either: a human voice took together-steps with a human face, allowing more than one voice/face, exactly as in the case of the sun and the moon. It is against this background that we must understand Aristotle's well-known remark that "spoken words represent mental experience and written words represent spoken words." If mental experience, even to some extent, had to be spoken, we were not as loose from our voices as we are today. If my mental experience can be silent in the world away from my face, it can be a *secret*, but if my mental experience cannot be silent in the world away from my face, another can put his or her face in my voice *to share* the experience—it is *between* us.

We began to think *with* our voices when we still spoke-together with each other and the world. The first thoughts were between us, *in the open*, neither inside nor outside. Jaynes' idea of "the bicameral mind" that preceded our "consciousness" turns on the possibility of our silent thoughts being "heard with the same experiential quality as externally produced sounds" (86). If indeed we did have such experience, we could no longer have been in the open; we would have had to make our places in the world in such a way that silent thoughts could be away from each of our faces in the world *alone* "with the same experiential quality as externally produced sound." Our thoughts would have already not been between us, but silent in the world away from our faces, in need of a new kind of speech that could carry them between us, as Vico put it, "from a distance."

It is about our lives in the open that Jaynes' remarks about sound are especially apt:

> Sound is a very special modality. We cannot handle it. We cannot push it away. We cannot turn our backs on it. We can close our eyes, hold our noses, withdraw from touch, refuse to taste. We cannot close our ears though we can partly muffle them. Sound is the least controllable of sense modalities. . . . (96–7)

And all the more so if I am simply in the open. Indeed, wasn't the posture of making a place in the open the ground for the original ascendancy of sound among the sense modalities? In the open I cannot gain a secret distance from a human voice. Hence the natural authority of a human voice, which is especially intimidating if, to boot, whatever personal-face one did have arose *through the voice of another*, as if only the other's face were possible, as if the world were able to be referred to a personal-face only if it was the other's face.

> Thus there appeared in the market place as many masks as there were persons (for *persona* properly means simply a mask) or as there were names. The names, which in the times of mute speech took the form of real words, must have been the family coat of arms, by which the families were found to be distinguished among the American Indians. And under the person or mask of the father of a family were concealed all his children and servants, and under the real name or emblem of a house were concealed all its agnates and gentiles. (Vico, 1984: §1033)

We cannot imagine being subject to such concealment, especially in its earliest form, though often as young persons—like those who cannot read because all words are immediately referred to another, a teacher, who "owns" them (Dennison; and part three, section four, below)—we do have experiences not all that different from our ancestors.

Let us say, then, that the plebians gradually came to demand the right to and of their own faces. In a letter from a subject to his king, King Ashurbanipal of Assyria, who died in 626 B.C., certainly in the age of heroes, if not of men, we find this claim: "My eyes are fixed upon the king, my lord" (Brucker, 29). Such a claim was possible only to the extent that the subject was still not loose from the king's face, or at least, was still compelled not to make anything of such looseness—the residue of a practice from former times that was destined to fade away even more. As we will discuss in detail in section two of part three below, it was to the king's face that everything and everyone must be referred: everything and everyone were *his*, "to sustain and represent . . . in his own person" (Vico, 1984: §1008). After the subject's proper adulation of the king, he went on to request that the king allow his son "to stand in the presence of the king, my lord." Hereby the son would become more of an individual, emerging from the other subjects who would remain relatively undifferentiated as far as having a *human* face was concerned. At one time such a practice—though obviously it was no longer what it was before—might well have been necessary *to share* the experience of a human face. Here is the true origin both of the natural authority of a human face, and of what Foucault (1979) called "ascending individuality," an ordering that survived long after the times we are discussing now. Only those with human faces and voices—literally—stood out as individuals, the others remaining more or less undifferentiated, depending upon how they were related to those individuals. (According to Thompson, women were the first to stand out in this way. Or was it only certain women, mothers, the original parents? In this regard perhaps, Thass-Thienemann reported that "many languages have no word for 'parents' because 'father' and 'mother' were too different to be united by one word" (143).) We can no longer imagine the need for such ascendancy in a whole community, let alone such ascendancy itself (though, again, perhaps as parents we do experience something like it in our *own* families).

The root of the archaic term 'ostend' is to-stretch-out-in-front-of-one's-face. This root makes especially good sense when we consider the early human voice as its paradigm, and sound as its sense modality. When voice and sound dominated our lives, we also believed, not surprisingly, that we were "all body" (Vico, 1984: §570):

> The truth is that none of the pagan sects believed that the human mind was completely [if at all] incorporeal. Consequently, they thought that everything produced by the mind was sense, i.e., that whatever the mind does, or undergoes, is *contact of bodies*. (Vico, 1982: 69 my italics)

A contact, as Levy-Bruhl discovered, that went well beyond what he would regard as "visible." Nevertheless, a simple conversation can serve as its paradigm: originally our mental experience involved evident co-making, the *touching-together* of bodies, in the open. Here, for Vico, "mind" or "consciousness" was "immersed and buried in the body," "hidden *in* men" (Vico, 1984: §331, 342). (Accordingly, "in all languages the greater part of the expressions relating to inanimate things are formed by metaphor from the human body and its parts and from the human senses and passions" (§405).) Then the mind was "submerged in the senses," as opposed to "abstracted from the senses," as it would be in the metaphysics to come in "the age of men" (§821). Only later, for Vico, did we emerge from our bodies to have our present mental experience, though that experience was still hidden, away from our faces in the world instead of "*in*" us. Let us now turn to the ground for our present mental experience.

VI. First Views

We have noted that at one point in our development we could read aloud but not silently, at least not as a normal practice:

> Normally, it seems an ancient Greek or Roman had to pronounce each syllable before he could understand a written word. The written letters informed his voice; then his voice informed his ear; and finally his ear, together with the muscular movements of his vocal organs, conveyed the messages to his brain. . . . (Stanford, 1)

But certainly the eyes were at work:

> Probably among the well-practiced readers their eyes, voices, ears, and minds received the message simultaneously. We know, in fact, that their eyes moved on ahead of the speaking voice, and we know from our best

> authority on education in the classical world, Quintilian, that this separation of faculties could be rather a strain on the ancient reader. Quintilian did not foresee that the quicker-moving eye would someday completely abandon the slower voice. (2)

Why not? We know that in ancient Egyptian writing, certain hieroglyphs, most notably the one referring to the writer, had faces that could be drawn in different directions to indicate the direction to read. Presumably the Egyptians could see the hieroglyph's face, but yet not read silently. How was this domination by voice and sound undermined? What is the difference between ear-reading and eye-reading?

To take advantage of the clue provided by the faces of Egyptian hieroglyphs, we must consider a figure that has faces drawn in opposite directions:

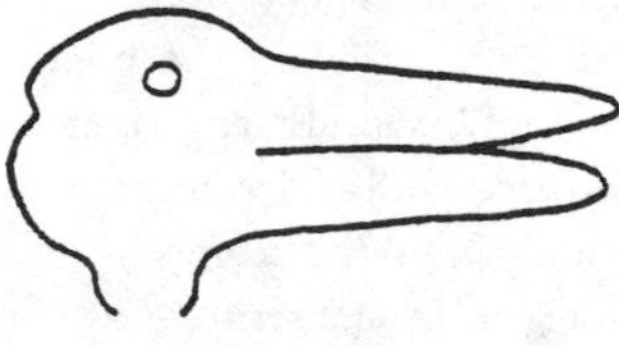

Figure 1

Light is reflected from the figure 1 in such a way that, should I put my face in it, I can refer the figure to my face in two quite different ways: either it faces left as a rabbit, or it faces right as a duck, though obviously it cannot do both at once. I use this left-right exclusion to switch between the rabbit and the duck, a switch that always occurs suddenly, at once.

The figure and the light reflected from the figure are equally ambiguous: away from my face, the figure can face left *or* right. My face is loose from the figure. The figure faces left or right in virtue of its being referred to my face, away from my face in the world, not in the world away from my face. When we say that the duck or the rabbit is in the figure away from my face, referring my face to the world, we do so because in both cases what is away from my face in the world is *also the same*, "congruent" (Wittgenstein, 1968: 199e): what the light *alone* can resolve. Strictly speaking, what I *see* remains the same, comprising what is away from my face in the world in virtue of my face being referred to the world. Hence—to jump ahead to the next section somewhat—a switch between *seeing* the duck and *seeing* the rabbit must involve a together-step of figure and face entirely lacking in the experience under discussion here.

In a purely visual experience of the figure I do not have *a view* of it. To be in view the figure must be referred to *a point of view*. If the orientation of my face in the world is referred to the world, my face, or rather, my face-place cannot be a point of view of the world; only if the orientation of my face in the world is referred to itself can my face-place be a point of view of the world. Both the bee and I can see the figure, but only I can view it as well. Indeed, I would require some considerable meditation simply to see the figure, as it immediately comes into my view facing right or left. To repeat, the switch between the figure facing *left* and the figure facing *right* must be referred to *a certain face* in the world: the duck or the rabbit comes into view *away from that face.* The root of the term 'distance' is to-stand-apart. When the duck or the rabbit comes into view, it *stands apart at once*; it is *at a distance at once.* Let us call this quality of a view "spatial ubiquity."

Now consider the figure 'now.' This figure also reflects light in which I can put my face. Are the figure and the light ambiguous? Are they ambiguous in exactly the same way as the duck-rabbit figure and the light reflected from it? Does this seem like an odd question? Perhaps long ago it would not have seemed so odd, when it was more commonplace to understand that words must face away from their beginning. Once again, both the figure and the light reflected from it are ambiguous in the world away from my face; only away from my face in the world does the figure face right as n-o-w. *N-o-w stands apart at once*: the figure *reads* n-o-w. The figure could also be in view facing left as w-o-n; or rather, it could *read* w-o-n. We are so accustomed to the figure facing right that we cannot, at least not without some considerable effort, have the figure facing left in view (let alone merely have the figure in sight), though numerous human beings around the world read just as automatically from right to left, for example, in the Hebrew language. *Eye-reading is an art of looseness, not of seeing alone.*

How does one see when ear-reading, without the world in view? That is, if I cannot refer the written figures to my face, how do I know the order to vocalize the figures? The answer can only be that the figures are referred to the world in some way. We must proceed very carefully here, because the very meaning of the faces of hieroglyphs is radically different if they must be referred to the world because the reader's face is itself referred to the world. The reader cannot just face the hieroglyph itself, independently of how the hieroglyph is placed in the world, as we do with our written figures. The reading cannot stand apart at once in this case; the reader *must speak to read*, for otherwise, of course, he or she would be a silent reader. The face of the hieroglyph can be thought of as a clue, referring the reader's face to *another aspect of the world.*

To make sense of this referral, let us begin with Thass-Thienemann's investigation of the origins of the relevant terms:

> "My pen is my plowshare," the German *Ackermann*, "plowman," said. One has no difficulty in understanding when Cicero wrote to Atticus: *Hoc litter-*

> ularum exaravi—"I plowed out this little letter." The pen "plows" the paper. In fact, this is the proper idiom in Hungarian. It is in accordance with these fantasies that the written lines are called "furrows." In Latin the lines were called *versus*, in English, *verse*, properly something "turned over," from the verb *verto, -ere, versum* "to turn." One can even observe that the "furrow-lines" were running from the left to the right and then from the right to the left in early literacy—a strange way of writing which must have had some motivation. Its Greek denotation gives the answer: it was called *boustrophedon*, properly "oxen-turn," equating the lines with the furrows of plowing. (387)

Strange? Not if the reader's face must be referred to the world in some way to know the order to vocalize the written figures. Notice that for a modern reader the feat of reading from left to right and then from right to left is nearly impossible: to have the written figures stand apart at once facing right and then facing left is, as already noted, no small achievement. But such an achievement was not involved in reading the "furrow-lines." The ancient Greek reader understood the order of vocalization by following the lines as if he or she were plowing a field. The ancient Egyptians also read from left to right or from right to left (or even from top to bottom, which could also be referred to the world or not). Notice that it does not matter whether or not such flexibility must be displayed in reading one series of written lines. Today we could not display it even if we were dealing with two separate series, each to be read in a different direction.

Given that the ancient Greek reader knew that the written lines were like furrows in a plowed field, he or she must still have needed to gain an orientation to those lines. (Think of the popular joke about people who cannot read today: they have the paper upside-down.) The orientation could be *exactly* that of a plowed field itself! For example, whenever the paper with written figures, flat on the earth or a suitable table, was aligned from bottom to top toward the west, the reader could read from south to north and then from north to south, as if the paper were a field to be plowed in this way. (The faces of the Egyptian hieroglyphs could also be aligned to the south so that the reader could read from south to north.) Moreover, the written figure to be vocalized first could be in the southwest corner of the paper, which corner (or figure) could also hold a clue by being suitably different from the other corner (or figures). (This practice should remind us of many old documents. We could easily imagine that originally the written paper was more like a map in the hands of an ancient reader. Note that our paper is often entirely blank save for the written figures.) Altars also tended to be built so that they were all aligned in the world in the same way. In ecclesiastical usage, east is the direction of the altar of a church looking from the nave. Hence a written paper properly placed on such an altar would be aligned exactly as the written paper above was aligned; any reading left to right would automatically be reading south to north, and vice versa. Was this one of the reasons for the constant placing of altars? Certainly it couldn't hurt an ancient ear-reader to be able to take such an alignment for granted.

To repeat, herein we are trying to understand how ancient readers could see well enough to read and write but not silently, not articulately. *They still could read and write*! Here was a degree of looseness not yet at the level we take so much for granted today, but nevertheless at a level that we must be very careful to appreciate as it deserves. Even before the advent of our looseness today, our language, no longer fully "mute" or "natural," was far more profound than we usually take it to be. What was fully "mute" or "natural" was no doubt constituted by the way our faces were immediately referred to the world, but between this extreme and the other one, that of fully "vulgar" or articulate language, our faces must have together-stepped with the written lines—the "furrow-lines"—exactly as they did with the spoken words, to constitute an experience of meaning that we cannot easily imagine, especially in its profundity. Do we really appreciate the level of achievement reached by our own children before they learn how to read and write silently? We tend almost to laugh when they draw letters facing in opposite directions in the same word (or draw things in the world without perspective.) But not too long ago at all, none of us could yet refer the world to our faces in such a way that a written figure could stand apart at once, *already read* one way or another. (We *became* fascinated with perspective.) We *invented* silent reading and writing, and gradually transposed our more or less naturally transparent language into one that increasingly depended on our *own* faces. Now we can easily forget that *originally we had no views*. We simply spoke and wrote without the world in view.

If I cannot have a view of written figures, I must depend on the temporal ordering of my speech to read, whereas if I can have a view of written figures, I can depend on the spatial ordering of my view to read. Then I also do not need elaborate safeguards against the ambiguities in the light reflected from the written figures, though our texts do contain certain clues, such as a capital on the first letter of each sentence and a period at the end. These conventions, like the faces of the hieroglyphs, are not voiced in the world away from our faces. Recall what it is like to listen to a foreign language spoken between native speakers: the speech seems continuous, run together in such a way that we cannot pick out even the sentences, let alone the words. Hence picking out the words from our voices must be compared to picking out the words from our writing: we cannot *hear* the former, exactly like we cannot *see* the latter. It is therefore helpful to introduce silence into speech when one speaks to a novice.

To understand the original quality of speech we cannot depend on our speech today. If we follow Jacques Derrida (1982: 21–5), for example, we will not be able to imagine speech that can fulfill its own promise, "to be *present* in and of itself": our writing does not "erase itself before the plenitude of living speech." But, Derrida notwithstanding, *originally it did do so*: the true origin of what Fernand de Saussure called the "thought-sound," the presence to itself which is at its extreme in a naturally transparent voice, the voice of *mimesis*, the voice that needs only to be *heard* to be understood. Today we do not *hear* any more than we *see*. We participate in conventional, not natural, transparency. The basis of the system of our written and spoken discourse is what Derrida called "archi-writing" (Culler, 171–2):

> Nothing, neither among the elements nor within the system, is anywhere ever simply present or absent. There are only, everywhere, differences and traces of traces. . . . The gram as *differance* [the element of archi-writing], then, is a structure and a movement no longer conceivable on the basis of presence/absence. *Differance* is the systematic play of differences, of the traces of differences, of the *spacing* by means of which elements are related to each other. (Derrida, 1982: 26–7)

Yes, conventional transparency is a matter of *spacing, away from a point of view*: no difference is more basic than that of left-right (even when it comes, as we will discuss before the end of this section, to the traces that reach through promises to connect the past and the future). The expression "in one ear and out the other" can serve as an epithet for speaking and listening with the world in view, as spacing subtly displaces the original temporality and silent background of speech (Merleau-Ponty, 1964b).

If a chant can be difficult to refer to one's own face, so too can silence. Both undermine all sense of sound coming and going, as one's face is referred to *one* chant, *one* silence. Ultimately one achieves, let us say, "temporal ubiquity," in which past, present, and future are not broken apart. Between temporal ubiquity and the other extreme of sound, when sound "goes in one ear and out the other," we experience degrees of temporal ubiquity. An archetypal life must enjoy some temporal ubiquity, and so too must a song or verse. The melody and rhythm require all the notes to be enfolded in a single whole theme so that past and future notes are not spaced apart from the present one as if they were going out of and into existence, respectively. (If they were, the present one "in the head" would be the only note in existence.) In chanting and meditation such enfolding can reach an extreme, overcoming spacing with temporal ubiquity.

The "right posture" of zazen, or zen meditation, permits "no idea of time or space," or alternatively, "no idea of place separate from time" (Suzuki, 29–30). A crucial aspect of the "right posture" is this:

> When you sit in the full lotus position, your left foot is on your right thigh, and your right foot is on your left thigh. . . . When I have the left foot on the right side of my body, and the right foot on the left side of my body, I do not know which is which. So either may be the left or the right side. (25–6)

If I do not know which is which, I do not have the world in view; I do not refer the world to my face. What do views have to do with an "idea of place separate from time?" Suppose I am once again facing the duck-rabbit figure. I can punctuate a certain temporal order by switching back and forth between the duck and the rabbit. When either suddenly comes into view, *another present* occurs, *in between* a past and a future. Such a present is the moment of a certain spatial ubiquity: at-a-distance-at-once-duck-or-rabbit. Hence, I cannot help but have an "idea of place separate from time." It is *away from that place* that I punctuate present moments: *that place is my point of view*. If in the "right posture" I do not make my place at a point of view, I must enjoy some temporal ubiquity.

As we will discuss in more detail in the next section, neither *seeing* nor *hearing* involves an idea of place separate from time. Some temporal ubiquity is evident unless a point of view intercedes, and then the view supports an idea of space—*the space of the world in view*—separate from time: what goes out of and into *that space* must go out of and into existence, the paradigms being *the past* and *the future*, respectively. Both our sense of history and our silent reading-writing have *the same origin*: the posture of making a place at a point of view!

As the original temporal ubiquity of voice was gradually eroded, in its place appeared the permanent mark, designed to last through a future broken apart from a past. What was automatic became the product of a technology:

> The verb *to write* originally denoted something quite different from the rather complacent activity which it came to mean in our age. The original meaning is still transparent in the closest related languages. The corresponding German word is *reissen, riss*, "to tear apart," from the former *rizan*, "to cut, tear, split"; . . . . In Greek the term *grapho*, "to write," carried similar implications. It meant "to encarve"; *graptus* means a cut on the skin. The Latin verb *scribo, -ere*, also denoted originally "to engrave, scratch in". . . . Material explanations are at hand. It is said that writing was originally epigraphy carved into hard material—stone, metal, wood—thus the letters had to be engraved by force or violence. (Thass-Thienemann, 386)

It is no accident that such writing began with the Bronze Age. But even before this time an earlier form of writing—as we might have expected given our interpretation of it above—was aimed at "cutting, tearing, and splitting" the earth:

> Again, a coat of arms was called by the Italians *insegna*, and ensign in the sense of a thing signifying, whence the Italian verb *insegnare*, to teach. They also called it *divisa*, device, because the ensigns were used as signs of the first division of the fields, which had previously been used in common by all mankind. The originally real terms [or boundary posts] of these fields later became the vocal terms of the scholastics. . . . The conclusion to be drawn . . . is that in the time of mute nations the great need answered by the ensigns was that for certainty of ownership. (Vico, 1984: §486–7)

Literally, the first writing paper was the earth, and gradually we made our marks of "division" on other material, until finally their origin was forgotten, in the seemingly "complacent" writing of today.

How could this happen? Before we dig deeper into the displacement of the original temporality and silent background of speech, it is important to note a connection between the technology of writing and the technology of agriculture. Here is Thompson's conclusion about the latter:

> This uneasy relationship between male and female, Enki and Ninhursag, water and stony ground, is the alluvial mud upon which Sumerian life is based [from the middle of the sixth millennium B. C.]. The instrument which has enabled man to change his relationship to the earth is the plow. A new technology has enabled man to take gardening away from women and turn it into agriculture; . . . the plow is under the majesty of Enki. (163)

Indeed, why not conclude, given all of our evidence, that originally no sharp distinction could have been drawn between plowing and writing, and that both were under the majesty of Enki? Then too Vico's story about the role of "men" fits into the picture, though it is not a pretty picture, as we say. As we will discuss in more detail in the concluding section of the book, to this day "the voices" of men and women differ exactly as this picture portrays them:

> The interplay between the responses is clear in that she, assuming connection, begins to explore the parameters of separation, while he, assuming separation, beings to explore the parameters of connection. But the primacy of separation or connection leads to different images of self and of relationships. . . . Perceiving relationships as primary rather than derived from separation, considering the interdependence of people's lives, she envisions "the way things are" and "the way things should be" as a web of interconnection where "everybody belongs to it and you all come from it." (Gilligan, 38, 57)

Here Thompson would no doubt have had us mourn "the lost religion of the Great Mother," which founded a life quite well described by this contemporary woman's voice. It is no wonder that women also claim that our language excludes their true voice!

VII. First Opacity

Let me tell a contemporary story, with a "primitive" twist. After driving somewhere on a sunny day, a father and his daughter return at night. The daughter asks, "Why didn't we drive home on the road with the sun shining on it?" Evidently she still lives in temporal ubiquity. For her, the road with the sun shining on it need not be recollected; it can be made "present again," as in the case of "primitive men" (Levy-Bruhl, quoted in Cazeneuve, 84). It is as if the road with the sun shining on it were hidden in the world away from her face, and she were able somehow to move to unhide it: the road still has the sun shining on it. But for her father the road with the sun shining on it is not in the world away from his face, hidden or unhidden: to have the sun shining on it, the road awaits *another* day, a future broken apart from a past though like it because the sun shines on the road *as it did once before.*

The father obviously does not have only one point of view during the drive with his daughter. But won't he act as if he were to have only one? We say: "I got a good view of the whole thing." The father has exercised his ability to use his face-place as a point of view to such an extent that he can "make" the required place: as he tries to explain matters to his daughter—and his explanations will fail *exactly* as explanations of left-right will fail—he will act as if he were to have a good view of the whole drive, with the events of the drive arranged one after the other away from his face in the world. His daughter, on the other hand, cannot "make" such places, let alone understand her father's explanations: her face still together-steps with the relevant events. To understand her together-steps, we must first understand her *seeing.*

The way the duck or the rabbit comes into view is not the way either could come into sight. A sudden switch between *seeing* the duck and *seeing* the rabbit would affect our faces as well, as if a together-step had been taken from duck/face to rabbit/face. We would think that something strange had happened, that the world had pulled a trick on us. This purely visual experience is hard for us to imagine, for it requires our faces not to be points of view, but to be referred to the world all along. James Gibson discovered the following law of the together-steps in seeing:

> Any movement of a point of observation that hides previously unhidden surfaces has an opposite movement that reveals them. Thus, the hidden and unhidden interchange. This is the *law of reversible occlusion* for locomotion in a cluttered habitat. It implies that after a sufficient sequence of reversible locomotions *all* surfaces will have been both hidden and unhidden. (193)

Given that the point of observation and what is hidden or unhidden always take together-steps, we might wonder how the world had conspired to hide the duck and unhide the rabbit even though our point of observation had remained the same. Or, we might wonder how the world had conspired to take the duck out of existence and to bring the rabbit into it:

> A surface that has no visual solid angle at any point of observation is neither hidden nor unhidden. It is out of existence, not out of sight. (193)

If the duck suddenly replaces the rabbit in sight, all the conspiring must be on the side of the world, but it can be on the side of us if the duck suddenly replaces the rabbit in view. This difference makes all the difference in the world!

For the visual world, indeed, it is not even appropriate to refer to a point of observation if thereby distance is understood to be reckoned "through the air." As Gibson took great pains to show, in the visual world distance should be reckoned "along the ground instead," by "the number of paces along the ground to a fixed object" (117). Hence, to contrast seeing with viewing, we must distinguish *a feet-place* from a face-place, or equivalently, *a standpoint* from a point of view. (The standard dictionary simply equates these two. Even the *OED* obscures their distinction: a standpoint is taken to be "a fixed point of standing; the position at which a person stands to view an object; a point of view.") At *a standpoint*, I have no sense of spatial ubiquity, of a place separate from time: I stand here, other things stand there, and we are *in touch through the ground*, along which we must *move* to close the distance between us. But at a point of view I have a sense of spatial ubiquity, of a place separate from time: other things are at a distance *at once, through the air*. I am, as it were, *swept off my feet into the space of the world in view*, no longer in touch with or through the ground; I have a sense, not of a distance to be moved through, but rather of distance separate from time.

The daughter is still on her feet, making places on the ground, at standpoints: she *sees* the world, and the road with the sun shining on it simply passes out of sight. But the father makes places in the air, at points of view: he *views* the world, and the road with the sun shining on it simply passes out of existence. We cannot *see* something pass out of existence, though something can suddenly pass out of sight. (Gibson concluded that "the perception of the environment [that is, *with our faces referred to the world*] is timeless and past-present-future distinctions are relevant only to awareness of self [that is, *with the world referred to our faces*]"(195).) Levy-Bruhl's "primitive men" live in the visual world, and he should have said, not that a certain portion of that world was "invisible," but that it was "unviewable."

The Hopi, who "are thought to be the custodians of the spiritual doctrine of traditional American Indians" (Boyd, 139), provide an adult version of the daughter's experience. Making places on the ground, at standpoints, they *see* the world, though not with the daughter's innocence.

> What happens at a distant village . . . can be known "here" only later. If it does not happen "at this place," it does not happen "at this time"; it happens at "that" place and at "that" time. Both the "here" happening and the "there" happening are in the objective, corresponding in general to our past, but the "there" happening is the more objectively distant, meaning from our standpoint [or rather, from our point of view], that it is further away in the past just as it is further away from us in space than the "here" happening. . . . [The] objective realm displaying its characteristic attribute of extension *stretches away from* the observer toward the unfathomable remoteness which is *both far away in space and long past in time*. (Whorf, 63, my italics)

A distant event may or may not be "here" in my future, but it *always* remains "there" in my future, *away from my face in the world*. Events do not pass out of existence, but neither do they merely pass out of sight: they are hidden away from my face in the world, not in the world away from my face, and so no ordinary movement along the ground can unhide them. (Perhaps an extraordinary or magical movement can, however.) We tend to have trouble imagining what this away-from quality is, for it involves both space and time, neither at a distance nor at once. Here is a degree of visual looseness that does not sweep us off our feet, though with it we could read and write easily, if not silently.

The Hopi, again, do not make places in the world away from which a distant event "exists at the same present moment" (Whorf, 63). (Even if I could not get a good view of a distant village, I could "make" a place away from which I could have gotten a good view of the whole thing. Obviously the Hopi do not "make" such places. Hence, spatial metaphor has no basis in Hopi experience: "the absence of such metaphor from Hopi speech is striking, as if on it had been laid the taboo teetotal" (146)!) Indeed, their "metaphysics does not raise the question" as to whether two distant events exist at the same present moment or not:

> DISTANCE includes what we call time in the sense of the temporal relation between events which have already happened. The Hopi conceive time and motion in the objective realm in a purely operational sense—a matter of the complexity and magnitude of operations connecting events—so that *the element of time is not separated from whatever element of space* enters into the operations. (Whorf, 63, my italics)

The Hopi are "frankly pragmatic": distance is a matter of *human effort*, referring us to the *action* required to *span* it. Thass-Thienemann found a similar foundation for our own terms of space and time:

> (1) The word *space* derives, through French, from the Latin *spatium*, which primarily denoted "room," a limited space. The word is a derivative of an old verb meaning "to span," "to stretch," like the verb *ex-tendo, -ere*, "to extend" from *ex*, "out," and *tendo, -ere*, "to stretch." It is used with the meaning "to stretch out," extend in space. In English a *stretch* still means a limited distance. "Space," as a "stretch," implies a dynamic action, an effort. (361)

> (2) The Latin term *tempus*, "time," originally denoted a "stretch," a short period filled with specific action. This concretistic idea of "time" will be better understood if one keeps in mind that the . . . idea "to stretch," which was also used for denoting "space," implies not only the limitation, but also the consequences of stretching. . . . thus the Greek *kairos* and the Latin *tempus* also mean "opportunity." This critical point of time is called the "nick of time" in English. . . . The verb *to nick* means to hit exactly the critical moment. All these instances, and many more in the other related languages, prove that the original concept of time denoted a short span, a unique opportunity for action. Its dynamic aspect is best pointed out by the term *moment*. It derives from the Latin *momentum*, originally *movimentum*, "movement." (370–2)

Notice that in both cases the terms do not refer to a distant event at all, but rather to a "stretch" of here-and-now, to an opening for action, which may well pass: "the word *past* derives from the Latin *passus*, 'step.' 'It *came to pass* at that time' (Genesis 38:1) is an obsolete formula, but it describes well the experience of someone walking along the way" (373). Yes, if *walking*, on my feet, at standpoints, *space and time are inseparable aspects of movement along the ground.*

If I am on the ground, my experience does not involve full temporal ubiquity because the future is broken apart from the past. But neither does my experience involve spatial ubiquity; as with the Hopi, I make "no attempt to distinguish present and past" (Whorf, 59). Originally, an opening for action, a moment, is *always dynamic*, a "stretch." (Notice that even the sudden coming into view of the duck or the rabbit is always already in need of recollection.) For a distant event to exist at "the same present moment," I must get apart from it at once, *away from a point of view*; it is there at once, *away from here*. It is not simply there, nor am I simply here. Gibson likened the simple "here-there" relationship to the "there-there" relationship because they both work the same way, through *mutual separation* along the ground. Herein hiding must be *co-hiding*, if hiding at all, for no place is favored as a reference point: everything is in the open in the visual world.

Gibson had this to say about a moving observer in the visual world:

> It is obvious that a motionless observer can see the world from a single fixed point of observation and can thus notice the perspectives of things. It is not

> so obvious but it is true that an observer who is moving about sees the world at *no* point of observation and thus, strictly speaking, *cannot* notice the perspectives of things. The implications are radical. Seeing the world at a traveling point of observation, over a long enough time for a sufficiently extended set of paths, begins to be perceiving the world at *all* points of observation, as if one could be everywhere at once. To be everywhere at once with nothing hidden is to be all-seeing, like God. Each object is seen from all sides, and each place is seen as connected to its neighbor. The world is *not* viewed in perspective. (197)

But neither is it viewed in perspective if I am at rest and my face is referred to the world, that is, if I *see* from a single fixed point of observation. As we will explain in detail in part two, in the visual world I am always embedded in an order of movement, working with a "stretch" of space and time, with no favored reference point; each place—including my place—is "connected to its neighbor," that is, *co-made with its neighbor*. It is extremely difficult for us to realize—and here Gibson obviously failed adequately to distinguish seeing from viewing—that we cannot *see* one thing *behind or in front of* another, exactly as we cannot *see* one thing *to the left of or to the right of* another. In this sense, *seeing is spaceless*. What Gibson really constructed was the space of the world in view; only that space is "everywhere at once." The result of traveling many paths in the visual world is the experience of the Hopi, not that of God. Only God really gets a good view of the whole thing!

So, let us suppose that we are in the visual world. What is the air like to us? Can it have near and far sides to it "at the same present moment?" *No*, between the near and the far sides is a distance that cannot be separate from time in this way. Moreover, I can touch the air with the whole surface of my body, save for my feet-place on the ground. As Gibson put it, "what one perceives [in the visual world] is an environment that surrounds one, that is in-the-round or solid, and that is all-of-a-piece" (195). So, as I am moving I can feel the air pass all over me. The air has viscosity, among many other qualities, all evidently associated with some sort of *body*. Like the ground, it affords us a sense of being *in touch with* the world.

Now let us suppose that we are in the space of the world in view, making our places at points of view. Does the air have near and far sides to it "at the same present moment?" *Yes*, it *must* do so! At one stroke, its bodily properties evidently vanish "into thin air," as we say. The air is so thin that a "stretch" of here-and-now is squeezed to its limit: *a point of view*. The root of the term 'transparent' is to-appear-or-be-visible-on-the-farther-side-of-or-beyond. This root makes sense only if on the near side is a face whose orientation in the world is referred to itself: *only away from a point of view is the air transparent*. Hence, viewing involves a transparency of spacing not at all involved in seeing. It is distance through the air away from a point of view that George

Berkeley labeled "invisible," mistaking, like Levy-Bruhl, the "unviewable" for the "invisible."

To understand how significant this difference is, first consider that originally the medium of voice must have been the air away from a standpoint, not away from of point of view. Sound is a modality of the air as some sort of body: we can be *in touch with* it all over our bodies, though especially through our ears (that is, through very fine hairs in a viscous medium deep in our ears). Sound surrounds us, never itself between us in a way that gives rise to an idea of distance separate from time. (P. F. Strawson has even argued that sound has "no intrinsic spatial characteristics" (57).) We are well advised to "keep our ears to the ground," however, for in the space of the world in view sound is not itself. To take only one of many examples, listen to another human being speak with his or her mouth in view, then not in view, then in view, until the difference is evident. When in view, the transparency of spacing calls upon the voice to stand apart at the mouth, *already read there*. It was *this* distance that liberated us from domination by sound and voice; the space of the world in view *housed* our voices. ("In one ear and out the other!") We no longer spoke-together with the world. *We had our own views!*

Here again is the Egyptian myth I quoted at the beginning of the introduction:

> Space only came into being when . . . the god of the air, Shu, parted the sky from the earth by stepping between them. Only then, as a result of his light-creating and space-creating invention, was there heaven above and earth below, back and front, left and right. . . . (Neumann, 1973: 108)

Is this not a perfect description of the advent of looseness, of making a place in the world at a point of view? Thass-Thienemann told us that "the 'sunlight' is, in primitive thinking, a very material substance, etheric, fluid, not heavy" (262). This is the light of the visual world, which, not surprisingly, retains a bodily quality exactly as the air of the visual world does. Imagine what it was like *to begin* to make places at points of view, with respect to which neither light nor air evidently retain their bodily qualities, especially that of extension never separate from time. (To take only one example again, think of shadows!) The near and the far sides, the earth and the sky, could finally stand apart at once, and in that space we found a wholly other light, the one that came to be the hypostasis for so much of our language about thinking:

> This clarity of thinking is supposed to break through the darkness of ignorance and prejudice and bring about the triumph of "insight" and "enlightenment" which is the specific human quality of existence. For this reason "to

> see the light" is the highest gift bestowed upon man. Sophocles' Ajax before committing suicide first says good-bye three times to light and only after that turns to his nearest kin, house, and home. Light is in the Greek conception the true house and home in which man lives. (Thass-Thienemann, 324)

No aspect of the visual world could serve alone as the basis for such experience. We *came* "to see the light."

As it is unlikely that we became loose all at once, or loose to the same extent, some of us must have had flashes of "insight" and "enlightenment," and only gradually did experience achieve a steady foundation for points of view. The way in which I can punctuate a temporal series by switching between the duck and the rabbit can easily be generalized to those cases in which I have a point of view of the flight of an arrow, of birds, or whatever. I can easily imagine, as did Zeno in fact, that a thing moving *in* the space of the world in view must somehow pass through an infinite number of moments, each one of which is immediately in need of recollection. But the crucial point here is that I *can* have experience of something *moving in space*. Such an experience, though ultimately replete with profound difficulties, as we will explain in detail in part two and in the first section of part three, cannot help but provide me with an ability that I do not have if I merely see. It is easy to believe that those who "scanned the flight of birds" in the *Iliad* were simply among the first to view the world. Obviously they would have known things in a way entirely mysterious to their visual fellows.

In this regard, Vico celebrated the fathers, who "were properly called divine in the sense of diviners, from *divinari*, to divine or predict" (1984: §381), and Homer celebrated Kalkhas Thestorides, who "came forward, wisest by far of all who scanned the flight of birds. He knew what was, what had been, and what would be, Kalkhas, who brought Akhaia's ships to Ilion by the diviner's gift Apollo gave him" (13–4—notice in this translation what appears to be a trace of Hopi tense construction). Apollo "is the shining god, he brings the hidden things into the open daylight" (Thass-Thienemann, 261). In this "mythos" aren't we celebrating our points of view and all that they make possible? After his drive with his daughter, the father can "make" a place away from which the events of the drive will be arranged one after the other as if he were to have a good view of the whole drive. If he can cast his imagination toward the past, he can also cast it toward the future, exercising the third of the original modes of memory: ingenuity or invention. Before these modes were broken apart, when some measure of temporal ubiquity still held sway, the work of "making" points of view would have easily qualified as successful divination, covering both past and future, as the *Iliad* reported.

Is the most ancient—and forgotten—dilemma whether to face the past or the future? Certainly this difference is fundamental to all languages (Thass-

Thienemann, 373). To face the past one does not break apart the present from the past (Hopi); to face the future one does not break apart the present from the future (Sartre). To do both, that is, not to break apart either the past or the future from the present—recall the figure of Janus, the Roman god of beginnings—is to be now, really, without past or future, as we were not so very long ago. (Evidently this experience can also be achieved through meditation.) But to be in between the past and the future, to break apart both the past and the future from the present, even for a moment always already in need of recollection, is the pivotal experience between facing the past and facing the future, for without it the latter would not be possible; that is, we could not have *turned* our faces toward the future, as we obviously had done, at least to some extent, in the times of Homer.

But what of Vico's "mythos" of the advent of "acts of human love?" Once again we must look to what points of view make possible. When I *see* another human being, we are separated by a distance that we need to cover, as we say. In a way that is terribly elusive today, the air already covers that distance, again as one of its bodily properties. When the god of the air, Shu, parts the heaven from the earth, he also uncovers us: now another human being stands apart from me at once, without any sense of a distance to cover, and vice versa. I need to hide to be "out of the sight of heaven." " 'The sun will detect it'. . . the German proverb says" (Thass-Thienemann, 262). (Do we also recognize here how the sun came to replace the moon as the crucial celestial reference at the advent of looseness?) We still do not engage in "acts of human love" in the light of day, as we say. In the light of day we must be covered; we wear clothes (though no doubt clothes were useful long before they were covers), and are mystified by how easily "primitive men" do not do so. But they are already covered by the air!

Alas, the irony: just as the skin is uncovered by Shu, so it becomes a cover itself.

> This implication is conspicuous in the word *hide* which means "skin," "pelt," and as a verb "to keep out of sight". . . . The skin is the mask behind which the true Self is hidden. In Latin *trahere pellem*, properly "to pull off the skin," means to take off the mask which conceals a person's faults. . . . The skin being such a protection of the Self against the outside world, to say "you get under my skin" is the expression of utmost irritation. (Thass-Thienemann, 213)

But the skin cannot play this privotal role at the boundary of outside-inside unless it is in the space of the world in view. Then at that boundary the view must stop: the skin is *opaque*. In the space of the world in view one thing gets in front of another thing *to hide it at once*. Here is the sense in which clothes are *covers*, for otherwise the air would already qualify as a cover. Indeed, here is the sense in which colors are *covers*: the root of the term 'color' is to-cover. Opacity is a matter of the color of things in the space of the world in view. Color stands apart at once, and no movement along the ground will eliminate it as movement will unhide what is hidden in the visual world. I must get under the skin—the color, the cover—to discover the true nature of a thing. Hence, as soon as the world is in view, not only *can* I question things, I *must* question things: *they are no longer merely in the open*—I *must* get inside them. The first natural opacity came along with the transparency of air, rupturing natural transparency, the transparency that requires no idea of place separate from time. As Vico noted, the term *minuere* originally meant division *and* diminution or corruption, and, according to the OED, in the Italian of about 1600 *divisare* meant divide or part asunder *and* devise or invent or discourse or think! These conjunctions are certainly no accident, though they may well be lost on us now after so many years of inquiry by division.

First level opacity has served as a challenge to us to discover a point of view that reveals the nature of the world. Consider "a book." Start with any written figure on a page of writing, and gradually widen the view until the written figures cannot all be held in view at once. Although each of us has a limit here, that limit does not stop any of us from "making" even wider views away from our faces in the world: the space of the world in view hereby houses "a book" as if we were able to read it at once—what Derrida called "the myth of a total reading." Not too long ago really, it became popular to imagine that from God's point of view the whole world is in view in the same way that "a book" can be in view. As we noted in the introduction about Galileo, he could not help but imagine "that great book," "the universe," as if it were written by God. Herein the point of view that has the whole world in view at a moment combines with the point of view that has the whole series of world events in view—one after the other or all at once?—to create the ultimate point of view. What a temptation to imagine that we could assume our own version of the ultimate point of view so that we could read the world too, as if the world were to stand apart at once, *already read* one way or another!

Such a point of view can no longer be said to be *in* the world. As we moved away from our original posture of co-making with the world we eventually reached "a place" at which co-making was itself ruptured, or rather, co-making was no longer *evident*. Witness the way a modern human being puts the problem as only a modern human being could put it: "Things appear as being outside of us whereas we should expect them to be in our interior" (quoted by Thass-Theinemann, 208). These "things"—colors, for example—are not regarded as co-made with things "outside of us"; what stands apart at once is somehow merely "subjective," not "objective." Nevertheless, even Plato believed that *viewing worked*, that what stood apart at once was "objective," co-made with things "outside of us." Let us turn to Plato.

VIII. Origin of Inquiry

Plato's theory of vision must be considered against the background of "early ages" in which "no differentiation was made between true being and outside appearance":

> The Greek thinking was conceived in the world of light, in the Appolonian visual world. It was not affected by the sophisticated discrimination between inside and outside, subject and object. It presupposed absolute confidence in the eyes and visual perception. They were convinced that existing things can be seen, thus immediately grasped by intuition. The Latin verb *intueri* is a compound of *in* and *tueri*, properly "to look at something." Seeing was a primary mental function of intuitive grasping of "being." The Greek language expressed this identification of "seeing" and "knowing" by a verb which means in the present *eidomai*, "appear," "shine," and in the past *oida*, "I know," properly "I saw." Thus, the Greek "knows" what he has "seen." (Thass-Thienemann, 146–7)

Plato continued this tradition, but only by distinguishing between "becoming" and "being": the eyes still worked to reveal existing things, but only those things in "becoming-space," not any true beings, the "unchanging" and "invisible" (1959: 53). Although visual perception was not true knowing, it was not "inside," somehow broken apart from "outside." Let us see how.

Numerous ancient Greeks believed that vision required some sort of emanation from the eye, a visual or ocular ray. Plato believed it too, but also made clear that the ray could not work as it needed to do if it were alone in the world:

> Such fire as has the property, not of burning, but of yielding a gentle light, they [the gods] contrived should become the proper body of each day. For the pure fire within us is akin to this, and they caused it to flow through the eyes. . . . Accordingly, whenever there is daylight round about, the visual current [the pure fire] issues forth, like to like, and coalesces with it and is formed into a single homogeneous body in direct line with the eyes, in whatever quarter the stream issuing from within strikes upon any object it encounters outside. So the whole, because of its homogeneity, is similarly affected and passes on the motions of anything it comes in contact with or that comes into contact with it, throughout the whole body, to the soul, and thus causes the sensation we call seeing. (1959: 42)

The term 'coalesces' is the translation for the Greek that literally means "*organic* fusion: a *growing* of things into one" (Aristotle, 1955: note to 438a). Hence, one stretches out in front of one's face, somehow growing together there with daylight, and the resulting growth can pass along the relevant motions of things. Elsewhere Plato stated more clearly just how crucial co-making is:

> As soon, then, as an eye and something else whose structure is adjusted to the eye come within range and give birth to the whiteness together with its cognate perception—things that would never come into existence if either of the two had approached anything else—then it is that, as the vision from the eyes and the whiteness from the thing pass in the space between, the eye becomes filled with vision and now sees, and becomes, not vision, but a seeing eye; while *the other parent* of the colour is saturated with whiteness and becomes, *on its side*, not whiteness, but a white thing, be it stock or stone or whatever else may chance to be so coloured. (1957: 47, my italics)

Here, without doubt, is an order of co-making in which the visible object, *on its side*, becomes a white thing. That which stretches out in front of one's face must also become transparent: the white thing appears or is visible on the farther side of or beyond it.

Although for Plato a white thing is only away from my face in the world, not in the world away from my face, I do stretch *to* that thing. He recognized both "the whiteness from the thing" which I put my face in, and also "a white thing" away from "a seeing eye": the thing itself comes, as it were, to wear its own color. The root of the term 'cognate' is to-be-born-together, so that "a white thing" and "a seeing eye" become at once: "a white thing" stands apart at once away from "a seeing eye." Does all this take place "in the space between?" Plato claimed so: "that *in* which all of them [what is of such and such a quality] are always coming to be, making their appearance and again vanishing out of it" (1959: 49). Hence, in our terms, "a seeing eye" is a point of view, "a white thing" is in view, and the space of the view is ubiquitous: a-white-thing-at-a-distance-at-once.

But exactly how is a together-birth of "a seeing eye" and "a white thing" possible? On the one hand, the visual current and daylight are *already* grown-together in such a way that the resulting body can be regarded as relatively ubiquitous as well as one and homogeneous. But, on the other hand, Plato still imagined that the body passed along motion, "the whiteness from the thing," in which one puts one's face. If the passing of motions is not at once itself, how is the together-birth to be explained? How does a "white thing" become "at the same present moment" away from "a seeing eye?"

Surely Plato's pupil Aristotle must have been concerned with this aspect of his teacher's theory, for on Aristotle's definition the light in which one puts one's face is itself a ubiquitous phenomenon:

> Now there is clearly something which is transparent, and by 'transparent' I mean what is visible, and yet not visible in itself but rather owing its visi-

> bility to the color of *something else*. . . . Neither air nor water is transparent because it is air or water; they are transparent because each of them has contained in it a certain substance which is the same in both and is also found in the eternal body which constitutes the uppermost shell of the physical Cosmos. Of this substance light is the activity—the activity of what is transparent so far forth as it has in it the determinate power of becoming transparent. . . . Light is as it were the proper color of what is transparent, and exists whenever the potentially transparent is excited to actuality by the influence of fire or something resembling 'the uppermost body'. (1947 (*De Anima*): 189)

Aristotle went on to criticize Empedocles for the theory that light travels but in its traveling is unobservable by us, for if it were to travel a long enough distance surely its traveling would be observable. No, for Aristotle light "is not a movement," and does not travel as sound or odor does. The actualization of transparency "may conceivably take place in a thing all at once": "the parts of media between a sensory organ and its object are not all affected at once—except in the case of Light" (1955 (*De Sensu*): 446b–447a). One of Aristotle's commentators, Alexander of Aphrodisias, interpreted these remarks as follows:

> Qualitative change is motion and occurs in time by gradual transition. But this is not the way in which a transparent medium receives light and color. It does not undergo a change but rather the situation is similar to someone becoming a right-hand neighbor without any motion or action on his part. Such is the turn which a transparent medium takes with regard to light and colors. . . . And just as the right-hand neighbor ceases to be on the right when the man on his left leaves his place, so does light disappear when the illuminating source is removed. (Quoted in Lindberg, 218)

Again we must recognize how crucial it is that the near and the far sides of a transparency exist "at the same present moment." The relationship of transparency between an eye and a thing, insofar as it is constituted by light, arises *at a distance at once*. In this respect, the quality of being in the light away from an eye in the world can be compared to the quality of being on the right away from an eye in the world. Yet in another respect the comparison fails: what I put my face in, according to Aristotle, is *always already* ubiquitous, in the world away from my face. Such a ubiquity can be thought of as playing a role quite like that of Plato's becoming-space. Not surprisingly, for Aristotle "light is neither fire nor any kind whatsoever of body" (1947: 189).

When it came to colors, however, Aristotle was as confused as Plato was. At one point he asked "whether the medium between the eye and its object is air or light" (1955: 438b)? He could not give up on motions being passed to the eye: "color sets in movement what is already actually transparent, for example, the air, and that, extending continuously from the object to the organ, sets the latter in movement." Yet Aristotle also believed that vision worked: "Whatever is visible is color and color is what lies upon what is in its own nature visible" (1947: 189–90). (Notice again how color covers things.) If he had settled on light as the medium of color, he would have had an easier time making sense of the visible object wearing its color away from a seeing eye in the world, and at times he seemed to agree: not only "in the case of Light" but also "in the case of seeing" "the parts of media are affected at once" (1955: 447A). Moreover, we can see fire "in darkness and in light" in such a way that we cannot help but conclude, at least in this special case of seeing, that the media is affected at once, "for it is just fire that makes what is potentially transparent actually transparent" (1947: 191). Alexander of Aphrodisias also claimed that *both* "light and color" deserve the same treatment. But then one must abandon air motions in favor of the ubiquity of "light and color," again as if a near side were not necessary. So how could Aristotle also define color "as the limit of the Translucent in determinately bounded body" (1955: 439a)? That is, how could it be such a limit—be the far side of a transparency—and yet share light's ubiquity? Without a movement being associated with color, how could one explain the existence of different colors? Then again, with a movement associated with color, how could one explain why the different colors stood apart at once?

Here the crucial point is that we began our modern inquiries into the nature of the world while we still believed our eyes, as we say. Color *stood apart* at once: *on its side* a thing wore the color that was away from our faces in that thing. Such a quality of experience was indispensable at the origin of inquiry. We began to inquire *with* experience of this quality, a quality still taken for granted in everyday life. (Centuries after Plato and Aristotle, David Hume prescribed some everyday experience as a cure for skepticism.) Our modern inquiries began with *the* world; although our points of view differed, we all had *the* world in view.

Consider a more specific example. Krupp referred to the work of Alexander Thom concerning the monument known as Kintraw in Scotland: "a partially buried boulder on the edge of the platform [a ledge on the very steep hillside] turned out to mark the exact place one must stand to see the winter solstice sun set behind Beinn Shiantaidh [a mountain] and watch it reappear with a momentary burst of light a few minutes later when it cut through the notch [between Beinn Shiantaidh and Beinn a Chaolais, also known as the Paps of Jura]" (35). This *momentary* burst is the key: "at Kintraw . . . it is the momentary beam of sunlight—a beam that can be seen on the right day only from the right place—that tells the bronze age skywatcher it is the solstice" (36). The skywatcher had a point of view away from which the light burst through the Paps of Jura; the sun stood apart at once, in view. Or again, the near and the far sides of the relevant transparency were "at the same present moment": at *the* moment the sun stood apart through the Paps of Jura the winter solstice began. We can almost hear the veteran skywatcher instructing an apprentice to assume the proper point of view, certainly not merely a standpoint: the apprentice *must face* the Paps of Jura *at the proper moment*. This is a lesson in posture, the posture of making a place at a point of view, with respect to which *the* world is in view.

However exactly we marked such moments—and, if only with huge stone monuments, we tried very hard to be accurate—we could not have questioned that we *had the ability* so to mark them. Had experience been of such quality that we went around challenging each other about having this ability—as opposed to challenging each other about how well we exercised this ability—our ancient skywatch would never have taken place. In some way, perhaps even a very subtle way, we have always known that this practice was not without problems, leading us to be careful. But the realization that no amount of care will achieve complete accuracy is the end of inquiry from a point of view, not the beginning. (Yes, I can be wrong about the proper moment of the winter solstice, but if I am shown to be wrong it must be by appeal to an experience *of the same quality*. Such experience is logically prior to being right or wrong about the world. Here René Descartes, but not Immanuel Kant, missed the significance of the qualitative similarity of illusory and non-illusory experience.) We must begin in innocence, the innocence of an everyday life based on a certain quality of experience. As Vico noted, philosophy came later (and not in all societies either): to question experience of this quality we must already be loose from it, and at first we were not loose from it, exactly as even earlier we were not loose from our voices. We can recognize this condition in Plato and Aristotle, who still believed their eyes.

Gibson made the following remarks about our experience:

> The optical information to specify the self, including the head, body, arms, and hands, accompanies the optical information to specify the environment. The two sources of information coexist. The one could not exist without the other. When a man sees the world, he sees his nose at the same time; or rather, the world and his nose are both specified and his awareness can shift. Which of the two he notices depends on his attitude. . . . The supposedly separate realms of the subjective and the objective are actually only poles of attention.(116)

But actually, these remarks must be directed to the times before we assumed the posture of making a place at a point of view. The crucial shift is not between "the poles of attention," but rather between *referring both to the world* and *referring both to the nose*. Only with the latter does the quality of experience required for modern inquiry arise: the world and the nose stand apart at once, the former constituting the far side, and the latter constituting the near side, of the transparency of spacing in views. Now the poles of attention are resolved in a way that leads us to divide them, whereas before, in the visual world, they were *evidently* involved in an order of co-making, as Gibson made clear. Light alone can resolve numerous ambiguities in voice and sound, but cannot resolve its own ambiguities. When we finally resolved them by making places at points of view, we could engage in the distinctively human form of inquiry, especially indebted to reading and writing in the space of the world in view.

What is at work in the space of the world in view is the modern form of attention, always already caught at a distance; an object of attention stands apart at once. Here is the aspect of visual perception that "is not a movement," as Aristotle might have put it. (The aspect also has an inside-out, as opposed to an outside-in, orientation, and has been regarded much as Plato regarded the ocular ray. Consider this remark by Rolling Thunder, a traditional American Indian medicine man: "Some people think seeing is just light coming in, but attention is a force that's emitted through the eyes" (Boyd, 225).) The irony is that this very form of attention provided us with the basis for its own "abstraction from the senses." Could we be *in* the space of the world in view and yet still have the quality of experience we obviously did have? Contact of bodies *in* that space seems to exclude, as Merleau-Ponty put it (1964a: 170), the "action at a distance and ubiquity that is the whole problem of vision," or rather, the whole problem of having the world in view. With the world in view, what would otherwise touch-together—through a little footwork at the least—is not together, but at a distance at once, and co-making *evidently* requires "action at a distance." Hence, the dilemma of Plato and Aristotle, a dilemma we still face today. The modern way out—and we can appreciate its lineage now—is (1) to put the space of the world in view into the world away from our faces, and (2) to take the view itself—*attention*—out of the world away from our faces: we do not stretch *to* the world. We abandon our bodies to space, in the world away from our faces, and then we make—or rather, to invoke our discussions of Vico, we "make"—our places elsewhere: "mind" or "consciousness." Hence, the advent of making a place in the world in the uniquely human way, as if we were not really in the world at all.

Part Two

The End of Inquiry

I. Body

At the beginning of part one, we could not as yet do justice to the claim that the light binds together the bee and the flower in such a way that it cannot be understood apart from them: we can no more break away the light from the bee and the flower than we can the bee and the flower from the light. In this part we will do justice to this claim. To begin with, however, we must study in some detail how its antithesis fits into the typical resolution of the dilemma faced by Plato and Aristotle. This resolution is best represented by the work of Newton. He clearly put the space of the world in view in the world away from our faces, and took the view itself out of the world away from our faces. Hence, the light is broken apart from the bee and the flower. The order of classical physics is not one of co-making.

More than a specific defense of an order of co-making, this part carries the application of the terms of posture to its natural limit. All along I have had the aim to remind us of the quality of our incarnation. Now I can say just how profound the aim will be: *I am my incarnation*. Although the typical resolution of the dilemma faced by Plato and Aristotle excludes any such aim, our adoption of it will help us to understand the history of inquiry about body. Our advantage will be precisely the work we have already done to develop the terms of posture; these terms provide a remarkably natural framework in which to tell the history of inquiry about body, including the most fundamental of such inquiries today, quantum mechanics. The claim held over from part one will simply be evident in our general conclusions about the history of inquiry about body.

No doubt many of us do not believe that we are body. I remember a number of my classes in which I asked the students whether they were body. Although I did not say anything at all about what I meant by the term 'body,' no one seemed terribly unclear about what was being asked of them. Most immediately declared that they were not body, or at least not entirely body. I declared that I was. Further conversation revealed some disagreement among those who initially agreed, but we still had the sense that we understood, if only tacitly, what the issue in question was. I propose to assume the same here, as I gradually work my way toward a complete account of this issue.

Now, more precisely, I am body, but only with a certain basic orientation. To get an initial idea of this orientation, consider our frequent yearnings to know other people as they know themselves. We wonder, for example, what it is like to look at the world through another person's eyes. As noted at the beginning of part one as well, a person's orientation is constituted by a face: I face out, from inside. We yearn to know another person from inside, inside out. *I am body inside out.*

Had Newton been a member of one of my classes, he would have declared that he was not body inside out:

> In bodies we see only their figures and colors, we hear only the sounds, we touch only their outward surfaces, we smell only the smells and taste the savors, but their inward substances are not to be known either by our senses or by any reflex act of our minds. (44)

Although unlike God we are not "utterly void of all body," our condition does not allow us to know body from inside. God alone knows the "inward substance" of body (44). Or again, as Vico put it (1982: 65), "the elements of natural things are outside us." Obviously on this account we cannot be body inside out; we can know body only from "outside."

So, Newton also believed that the seat of his experience of body must somehow elude the terms of analysis demanded by body itself:

> Every soul that has perception is, though in different times and in different organs of sense and motion, still the same indivisible person. There are given successive parts in duration, coexistent parts in space, but neither the one nor the other in the person of a man or his thinking principle. (43)

Although he disagreed with Newton on so many fundamental matters, Gottfried Leibniz agreed on this one:

> It must be confessed, however, that Perception, and that which depends upon it, are inexplicable by mechanical causes, that is, by figures and motions. Supposing that there were a machine whose structure produces thought, sensation, and perception, we could conceive of it as increased in size with the same proportions until one was able to enter into its interior, as he would into a mill. Now, on going into it he would find only pieces

> working together upon one another, but never would he find anything to explain Perception. (457)

Again, the experience of body is from "outside" body, even when we are imagined to *be* body. Though curiously self-defeating, this feat of imagination has held us captive to the present day. Opponents still tend to argue in its terms, save for the relativity of the magnitude of the "pieces."

Although, as we noted in the introduction, the kind of account of "Perception" given by Newton, Leibniz, and Vico can be traced back to Galileo, it was Descartes who made the crucial break with Plato by removing sensual experience from the body (Rorty). Consider the term 'yellow.' Descartes took the idea of yellow to be innate, tracing its innateness to the faculty of thinking, to the "inward source" of thinking, of which the idea of yellow is only an aspect. Outside this faculty, we find only movements of body but nothing at all that resembles yellow; yellow is that which arises in us as a result of these movements. In the language of a standard dictionary today, Descartes was making the distinction between the sensation of yellow—"a conscious sense impression"—and the series of events that leads to that sensation. Relative to this series we understand the term 'yellow' to refer either to the bodily property of reflecting light of a certain wavelength, or to that light itself, or perhaps even to our bodily response to stimulation by that light. All of these events are taken to be broken apart from the yellow of the faculty of thinking, let us say, the "inward" yellow.

What does this mean? Well, again, a standard dictionary will give two senses of the term 'yellow': (1) light of a certain wavelength or the property of reflecting such a light, and (2) "the color of gold, butter, or ripe lemons." Sense (1) we encountered at the beginning of part one in our discussion of the orientation of bees in the world:

> Poppies are red flowers that are frequented by bees. This exception has an amazing explanation. *Poppies reflect the ultraviolet rays of sunlight*. It is possible to show by suitable experiments that bees trained to the blossoms of the poppy are in fact recognizing *the reflected ultraviolet light*. We cannot perceive this light, and we see only red. The bees, on the other hand, cannot perceive the red; they see only ultraviolet. (Frisch, 1976: 18, my italics)

Is there also a sense (2) of the term 'ultraviolet' for bees: "the color of poppies?" Let us answer, sense (1) suffices for the bee; the ultraviolet color of poppies is always already in the world away from a bee's face. This answer also accords with an outside-in analysis of the color of poppies to bees, but only to the extent that the poppies are attributed the property of reflecting light of the ultraviolet wavelength. In the world away from a human face the relevant property of the poppy is to reflect light of the red wavelength. At this level of analysis, the poppy enters into *two* relations: one to a bee and one to a human being.

But what of sense (2) of 'yellow' for a human being? Isn't it ostensive? Isn't it asking me to stretch-facing gold, butter, or ripe lemons? Aren't gold, butter, and ripe lemons being referred to my face? Obviously yes, for I cannot use sense (2) unless I do face them, or equivalently, unless I have already faced them. More precisely, I must *put* my face *in* yellow in sense (1) in order to understand yellow in sense (2). But then yellow in sense (2) cannot be the light I put my face in; it must be *away from* my face *in* that light. Here again we encounter the difference between a face whose orientation is referred to the world and a face whose orientation is referred to itself: unlike the bee, I can *put* my face *in* the light in sense (1). The bee always already has its face in that light; no ostensive opening allows the bee to question the poppy rather than automatically to extend its proboscis. My face is loose from the light in which I put it in such a way that sense (2) of the term 'yellow' is required to understand my reference of gold, butter, or ripe lemons to my face. Why not say, then, that Descartes' "inward" yellow is simply yellow in sense (2): away from a human face in the world but not in the world away from a human face? We can still claim, as he did, that we bring such yellow to the world, exactly as we bring left-right to the world. No verbal tautology can define either one; neither is to be found among what Vico would have thought of as "natural relations": early human language, again, was "a mute language of signs and physical objects having natural relations to the ideas they [the early human beings] wished to express" (1984: §32).

Accordingly, sense (2) of the term 'ultraviolet' does not exist for a human being. If I put my face in ultraviolet in sense (1), no ultraviolet will be away from my face in the world. But then again, does ultraviolet in sense (1) exist for a human being either? Does ultraviolet exist in the world away from my face, or only outside my face, broken apart from it? Let us opt for the latter, so that the similarity of the bee's face and my face is that neither of them can be said to wait to be stimulated by light in sense (1) (that is, to use Frisch's term, by light they can "see"). Here is the level at which we remark about the range of sensitivity of different eyes, though we must be careful to add that our remarks are no support for an outside-in analysis of a stimulus to a face, on which we must allow both an ultraviolet stimulus to my face and a red stimulus to a bee's face.

Let us recall once again the recourse to facing north in order to understand left-right. The real problem with such recourse is that anyone who can follow the instruction to face north already knows what left-right is. We can see this problem when a child is trying to learn left-right. No one can teach the child: until the child has a face whose orientation can be referred to itself any teaching will be unintelligible, but once the child has such a face any teaching will be unnecessary. Left-right—though not 'left-right'—comes along *with* a face whose orientation can be referred to itself. Even if Descartes' way of putting it is not apt, he can be thought of as making the same point about yellow. Yellow—though not 'yellow'—comes along *with* a face whose orientation can be referred to itself.

We are unique in the sense that we are so loose from the world and its "natural relations" that we can live our lives in such terms as 'left,' 'right,' and 'yellow,' which constitute the foundation for all that we hold so dear. Yet again we are so loose that we appear to be broken apart from the world and its "natural relations," allowing Descartes to divide the faculty of thinking from the body, a division that has come to be a paradigm of a world in pieces. He simply pursued an outside-in analysis of his face to such an extent that the yellow away from his face was broken apart from the world altogether; he failed to appreciate, we can say, that the yellow was still *away from his face in the world*. One of the special tasks of part two is to do justice to this conclusion: even the yellow we can be said to bring the world cannot be broken apart from the world.

The success of the Cartesian position of breaking us apart from the world is all the more remarkable given its long history of difficulty in explaining itself: from "outside" body, how do I *know* that my experience is actually of body? Indeed, what *is* body? The resolution of these profound questions, I stipulate, requires that a person be body inside out. As with any stipulation its merits will become obvious only after a journey, following the progress of inquiry about body over most of its history. This stipulation also differs from others when it comes to an explanation for the need to make it: not only does it assume absolutely nothing about body—in which case it is, as Einstein would have put it, neither a supposition nor a hypothesis about the nature of body—but also the use of the term 'body' must gradually undergo a metamorphosis as the journey unfolds, so that at each stage of the journey the stipulation will take on a new meaning. At the first stage, that of classical physics, the stipulation hardly encourages understanding, let alone belief. Only at the last stage, that of quantum mechanics, will its actual force become evident.

I do not wish to seem mysterious. The journey herein must simply be thought of in the context of discovery, not of justification. As numerous scientists and philosophers have learned—or should have learned—the rivals to the above stipulation are no more open to final justification than it is. Hence, today we still engage in a lively debate about our being body. For the purpose of edification, at any rate, the problem is not so much to resolve the current debate as to recast it; it is still cast in terms that preserve the structure of the original debate. Overlooked, again, are precisely the terms of posture. If we *begin* with the terms of posture and the stipulation they make possible, what do we *discover* about our being body? We aim to "prevent conversation from degenerating into inquiry, into an exchange of views" (Rorty, 372), such as the exchange in the current debate about our being body.

Certainly my life, especially as a philosopher, has been dominated by a tendency not to appreciate body. Philosophy began, so Vico believed, only when we emerged from the order of "all body." Philosophy and eventually science were captivated by an outside-in orientation to body that captivated me as well. But if we tell the story of our inquiry about body in terms of the additional inside-out/outside-in difference which we have only begun to explore so far, the story not only makes more sense. We also open doors to new ways of understanding how body itself can perform the work of experience; namely, the new theory of how our nervous system works that I referred to in the introduction.

On the above stipulation, then, whatever the level of organization of body, I am body at that level. Although I am body from inside, I can be studied from outside. Or again, if I am approached from outside, the various levels of organization of body apply to me without circumscribing me. Newton also thought that he escaped this circumscription, but obviously not by being body from inside. I aim to show how to study body from inside, so that at each level of organization of body, at least at each level beyond that of classical physics, we may gain the additional insights required for a complete account of body and our experience of body. We will learn that the captivating posture that led us to believe that we were somehow "outside" of body—the posture of making a place at a point of view—is itself an aspect of an order of co-making of body.

II. Classical Analysis: Newton

"Place," said Newton (18), "is a part of space which a body takes up, and motion is the translation of a body from one place into another." Or again, it is body that takes up a place and moves from one place into another. The space of these places may be absolute or relative, and the former is defined "without relation to anything external" (17–8). An absolute place exists in itself, independent of whatever body takes it up; it is always already waiting for its occupant. Hence, places must stand apart at once. Newton put it thus: "every particle of space is *always*" (43).

Like the "coexistent" parts of space, "the successive parts of duration" may be absolute or relative, and the former is again defined "without relation to anything external" (17). A moment reaches through all places exactly as a place reaches through all moments: "every indivisible moment of duration," said Newton (43), "is *everywhere*." (A classical analysis tends to spatialize time, as if absolute moments were points on a line, each one always already waiting for its occupant exactly as an absolute place does so). Although God is neither space nor time, he "constitutes" both "by existing always and everywhere" (43). So, too, must he be responsible for the ultimate parts of body (178).

Now, even though we do not exist always and everywhere, we can make a place at a point of view. The space of the world in view has precisely the characteristics of relative space, and the moment that we can mark away from a point of view, in between the past and the future, immediately in need of recollection, has precisely the characteristics of a relative, indivisible moment. Moreover, if these relative places and moments are broken apart from points of view and put into the world away from our faces, they are easily imagined to exist "without relation to anything external." (Recall our discussion at the beginning of part one about the frog and the insect hung before it. A classical analysis tends to neglect the difference between being in the world and being in the world *away from our faces*, or equivalently, the difference between being broken apart from our faces *altogether* and being broken apart from points of view *alone*. It took Einstein finally to understand this mistake in physics.) Let us then say—and we will have occasion to add more reasons—that a classical physicist makes a place at a point of view, and that the characteristics of the places and moments he or she marks away from that point of view constitute the paradigm of absolute space and time.

Newton believed as well that the laws of motion apply to body in absolute space and time, *the* inertial frame of reference, in which a body remains at rest or in uniform motion (with no acceleration) unless some external force leads to a change in its motion. The aim here is not to allow either an accelerating or a rotating frame of reference. All other frames are "inertial," moving uniformly with respect to absolute space and time, in which a body is either absolutely at rest or absolutely in motion. What is at rest or in motion in our view may very well not be so in *the* inertial frame. Hence, we employ a preferred reference frame—or reference body, as we will see in a moment—for determining the velocity of a phenomenon such as light.

The notion of a body taking up or filling a part of space is difficult to understand from outside that body. How do I know anything at all about such filling if I am utterly excluded from it? I demonstrate the problem to my students by breaking a piece of chalk to display the filling: more chalk, like that of the surface of the original piece. But what happens when the chalk can no longer be broken?

To study body from outside we must depend upon the laws of division of body. If a body is a composite of other bodies, we divide the body into its parts to study those parts: their situation, relative motion, and so on. We come to know the original body from inside only to the extent that we can divide it, though these divisions are already inherent in the original body. Were they not inherent, the body would not be composite, but simple; it would have no parts for us to divide, and we would be forever excluded from its inside. The simple bodies we study only from outside: we try to account for every event in terms of these bodies, which unite and divide through their motion.

The laws of motion are posed in mathematical terms that turn on points of body: parts of space without size or shape, extentionless, yet assigned a mass. We are asked to imagine body in general to be constituted by these points of body. A popular introductory physics text refers to a point of body as "an idealized body called a *particle*," and the motion of a body that takes up a part of space is discussed in terms of the motion of "every particle of the body" (Resnick and Halliday, 30, 215). In the background here must be the notions of an indivisible moment and an indivisible place; a body is supposed to be localized in terms of such moments and places. Indeed, in the new classical mathematics—the calculus, or "the method of fluxions," as Newton called it—we understand an integral over space and time by thinking of a certain summation of intervals of space and time as they approach indivisibility: the limit is an infinitely divided space and time, composed of indivisible places and moments respectively.

We needed a way to study body without really being in a position to know the "inward substance" of body. Why not assume that body has no inside, that is, is constituted by points of body that literally have no inside? We imagine, in effect, that body shares with space its infinite division, thereby requiring the ultimate parts of body not to take up space at all. (As we will see in a moment, only Leibniz took this feat of imagination seriously at the time, even though, in later hands, it would turn out to have some sense.) The insight to work in these "ideal" terms made possible the discovery of classical physics and mathematics (Hall). Only some two centuries later—indeed, some three centuries after Galileo took the first steps—did we come to understand how to get inside Newtonian bodies.

Getting inside Newtonian bodies was only one of our problems, however. Moving between them was equally problematic. That is, if body is what takes up a place and moves from one place into another, and if two bodies are not in contact, then moving between them is possible only if, so to speak, we hitch a ride on a third body that moves from the one to the other—a case of indirect contact. Space in itself is an order of exclusion, as one place is absolutely outside of every other one. So, too, a body at one place—think especially of a point of body—must be absolutely outside of every other body: two bodies cannot take up the same part of space. Nothing else but another body (or bodies) exists in space to provide a means of communication between two bodies in different places.

Again, body in motion in space and time *is all there is*. All energy is kinetic; communication between bodies must involve a transfer of kinetic energy according to certain universal laws. The paradigm of this communication is by contact, as in the case of billiard balls that Hume made famous. Perhaps this paradigm does not seem too limiting at first glance. But pause for a moment to think of all the cases of bodies that are obviously divided in our view of them, yet that are equally obviously in communication as if they were somehow involved in a transfer of energy. Think, for example, of a magnet at work.

I know from years of discussing the matter with my students that few people who study classical physics realize the significance of the equation 'F = ma.' What does the equation tell us that force is? The acceleration of body, period! (I do not mean to claim here either that Newton actually thought in terms of this equation, or that the equation is not controversial. See Dijksterhuis, 469–77.) To continue our example, an object will accelerate toward a magnet, and a special form of the general equation 'F=ma' will express exactly how the acceleration will occur. We tend to imagine, again, that the magnet and the accelerating object are not in contact, not even indirectly through other bodies. But our imagination aside, such a lack of contact is not possible: an explanation must exist in terms of bodies in contact, transferring kinetic energy from one to another. No force can exist except as embodied in such a chain.

Hence, Newton faced another special dilemma when he came to explain gravity. His predecessors had been preoccupied with an explanation for what otherwise appeared to be action at a distance. Certainly as far as our senses can testify, the earth and a ball in flight above it are not obviously in contact through other bodies. The ball simply seems to fall to the earth, as if pulled there, at a distance. Again, bodies do seem to behave as if they were able to act on each other at a distance, instantaneously.

Newton dared to formulate a law of universal gravitation between bodies that does not, at least not in itself, entail any intervening communication by contact between these bodies. Many of his contemporaries declared the law to be "occult." All appearances to the contrary, gravitation between two bodies must be explained by the direct contact of other bodies, the cumulative effect of which is the motion of the two bodies. Newton made his breakthrough, however, by setting aside any immediate concern for such an explanation of gravity:

> That gravity should be innate, inherent, and essential to matter, so that one body may act upon another at a distance through a *vacuum*, without the mediation of anything else, by and through which their action and force may be conveyed from one to another, is to me so great an absurdity that I believe no man who has in philosophical matters a competent faculty of thinking can ever fall into it. Gravity must be caused by an agent acting constantly according to certain laws, but whether this agent be material or immaterial I have left to the consideration of my readers. (54)

The reference to an immaterial agent—no doubt Newton had God in mind—shows how profoundly Newton's law of universal gravitation does not entail the appropriate material agent, yet Newton himself never tired of providing an explanation in terms of just such an agent.

All connections between bodies must be *constructed* through chains of bodies in contact. No connections are simply given; they must be external to body, not internal to body, not "innate, inherent, and essential to matter." Inasmuch as Newton's law of universal gravitation does not exclude the possibility that connections between bodies are simply given, these connections may not be constructed at all, let alone by chains of body in contact. In Newton's day, as already noted, an unconstructed connection between bodies was called "occult," a regression to the days of Aristotle and his doctrine of natural places. An unconstructed connection must be internal to body, not external to body, as if a body were to have an affinity for a certain place, to be drawn there without any intervening connection, material or immaterial.

On Aristotle's view, a substrate, in itself formless, takes the form of being at a place and a time. The substrate is "matter," but in order to be what we normally call "a body" matter must take the form of that body, say, the form of a tree. Like the form of a tree, places and times are forms among other forms which do not exist in or by themselves, but only as the forms of matter. Indeed, each of the four elements Aristotle took to compose the universe—earth, water, air, or fire—is a certain form of matter, or alternatively, a certain actualized potential of matter. The natural place of earth, for example, is at the center of the universe; in being drawn toward the center, earth is simply actualizing its potential to be in its natural place. It is precisely this view that Newton ridiculed in the passage quoted just above.

In the case of light, as we noted in the last section of part one, Aristotle's matter has a potential to take the form of being transparent in such a way that "the parts of media between a sensory organ and its object are affected at once." Newton could not tolerate instantaneous external connections any more than internal connections, though he agreed with Aristotle in putting some sort of standing-apart-at-once in the world away from our faces. Aristotle simply did not accept a Newtonian notion of place; an Aristotelian notion of place presupposes a maker of that place (Furley), and thereby violates Newton's condition of being defined "without relation to anything external." Hence matter, the substrate, plays the role of, so to speak, the arena of Aristotelian physics: we study its potentials, the relationships between them, and so on. For Newton, on the other hand, the arena of physics is the absence of all such substrate, an emptiness without the potential to be or do anything at all—absolute space.

So, again, on the classical analysis two bodies at rest with respect to each other are not connected in such a way that they will move together or apart without some third body or bodies to provide the required connection. These third bodies were called "ether." For Newton light also involved the motion of some "subtle" body in space and time, as opposed to the instantaneous, immaterial connection provided by light on Aristotle's theory. Although the latter connection can be said to be constructed, the principles of its construction are decidedly not classical. On the classical analysis motion of body in space has duration.

We must stress the corpuscular aspects of Newton's theory of light, even though his "rays" of light could combine and undergo "fits" in ways that indicate wave-like properties as well. Until the dawn of this century, waves of light were always presupposed to work through some sort of bodily medium, an ether, in which they propagated in the continuous fashion of air or water waves. The strength of this presupposition is remarkable; even after the concept of a field had become essential to the formulation of physical laws, the presupposition still held sway. A field is a function of space and time coordinates that assigns a value of the field for each such pair of coordinates. For many years after their invention, fields were regarded as having to be *of* some sort of body.

Light will move at its natural velocity, its velocity of propagation through the ether, only if the ether is absolutely at rest. The ether absolutely at rest is the body to which we refer the propagation of light. Should the ether be absolutely in motion, light will move slower or faster depending on which way it is moving relative to the ether. We should be able in principle, if not easily in practice, to detect such differences, and if we had succeeded the etherial theory of light would have been vindicated—hence the famous Michelson-Morley experiment. Its failure near the dawn of this century set the scene for Einstein's special theory of relativity. Not only would we give up on an ether; we would also rethink what we meant by space, time, and body.

But before proceeding to discuss the special theory, we need to go into more detail about the classical analysis. We have noted the difference between absolute motion or rest and relative motion or rest. The problem of detecting this difference is profound, as Newton himself realized. The problem also belongs to a family of closely associated problems. The laws of physics were supposed to apply to body in itself, rather than merely to body as it appears in experience of body:

> Instead of absolute places and motions, we use relative ones, and that without any inconvenience in common affairs; but in philosophical disquisitions [physics included], we ought to abstract from our senses and consider things themselves, distinct from what are only sensible measures of them. (Newton, 20)

What is crucial to us here, as noted in the introduction as well, is this example: although body in itself does move, it has no color. Like the other proper objects of the senses, color is not objective, but subjective, merely the way in which body appears in sensual experience of body—hence the distinction between primary (objective) and secondary (subjective) properties of body, and the problems of sorting out which is which.

So, what are the events relevant to a visual experience? To begin with, said Newton (77), "all natural bodies are variously qualified to reflect one sort of light in greater plenty than another." The reflected light must carry the information required to resolve a visual experience. Here Newton was confined by his understanding of light and natural bodies:

> The ends of the capillamenta of the optic nerve, which front or face the retina, being such refracting surfaces, when the rays impinge upon them, they must there excite these vibrations, which vibrations (like those of sound in a trumpet) will run along the aqueous pores or crystalline pith of the capillamenta, through the optic nerve. (97–8)

A certain kind of vibration carries the required information. Elsewhere Newton expanded upon this idea:

> If when we look but with one eye it be asked why objects appear thus and thus situated one to another, the answer would be because they really are so situated among themselves and make their pictures in the retina so situated one to another as they are; and those pictures transmit motional pictures into the sensorium in the same situation. (102)

And when we use two eyes the respective motional pictures "come together and are coincident," at least normally. Hence, the series of events that leads to the resolution of a visual experience begins with the reflection of light from some natural body and follows that light to the eyes, where the constitution of the eyes and optic nerves allows them to transmit the appropriate "motional pictures" to the "sensorium."

What next? "By the situation of those motional pictures the soul judges of the situation of things without," said Newton (102). He made a similar remark about color (98): the appropriate vibrations run through the optic nerve "into the sensorium (which light itself cannot do), and there affect the sense with various colors according to their bigness and mixture." (The relevant series of events is illustrated in figure 2 below.) Still, in the case of color, it is hard to imagine how the motional pictures carry the information required by the soul. Certainly Newton was clearer about the primary quality of spatial situation than about the secondary quality of color. A cornerstone of the very distinction between primary and secondary qualities was the ease of imagining how motional pictures could be used in the

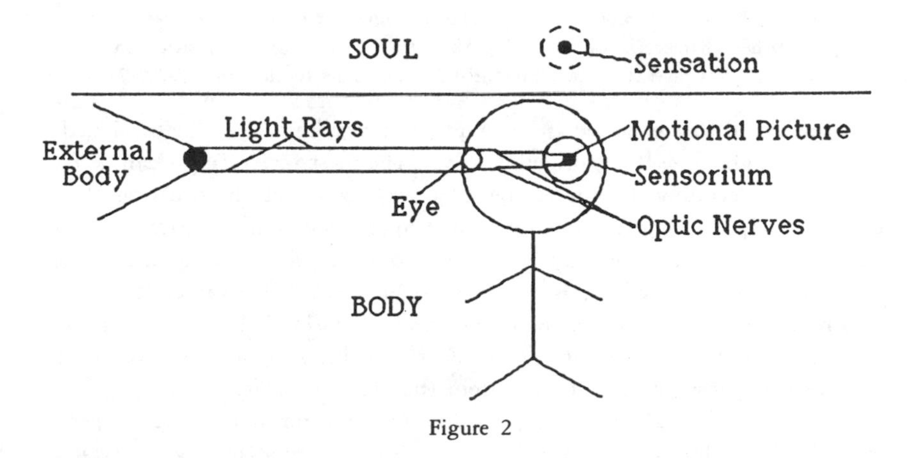

Figure 2

former case, though in both cases we have the same difficulty in imagining how the soul has access to the pictures.

Unlike God, who "sees the things themselves by their immediate presence to himself," we behold their "images":

> Is not the sensory of animals that place to which the sensitive substance is present and into which the sensible species of things are carried through the nerves and brain, that there they may be perceived by their immediate presence to that substance? (156)

Given Newton's judgment of the spatial organization of the motional pictures immediately present to his own soul, he must have thought of the soul as having *a point of view* of those pictures exactly like the point of view we would unthinkingly take ourselves to have of "things without." The elements of the relevant motional pictures must stand apart at once so as to represent things without "as they are" (102)—hence the primary quality of the space of the world in view. Only on this basis could he make the distinction between primary and secondary qualities in general.

Among the few classical philosophers-scientists to distrust this distinction, Berkeley still believed that, were there external bodies, colors could not be inherent in them:

> In case colors were real properties or affections inherent in external bodies, they could admit of no alteration, without some change wrought in the very bodies themselves; but is it not evident from what has been said, that upon the use of microscopes, upon a change happening in the humors of the eye, or a variation of distance, without any manner of real alteration in the thing

> itself, the colors of any object are either changed, or totally disappear? Nay, all other characteristics remaining the same, change but the situation of some objects, and they shall present different colors to the eye. (1979: 21)

Newton would have endorsed this statement, a paradigm of the classical analysis. Although we will not discuss it as yet, the most interesting example for us herein is certainly the microscope. One way to change the situation of an external body is to put it in the target of a microscope. On a classical analysis the body will not undergo any "real alteration." Another way to change the situation of an external body is to put it, so to speak, in the target of an eye. Again, the body will not undergo any "real alteration." In both cases, we must be able to assign definite properties to the body in isolation even from the observing instrument, let alone from the observation itself.

The classical analysis of sensation *ends* with the terms of sensual experience. First the things of the world must be described in a language especially designed for that purpose, the language of classical mathematics; we apply this language to each stage in the process of sensation, as we work our way toward the sensation itself. Each stage must be imagined to exclude the very qualities of sensual experience until that experience is actually brought about in the end. Among the excluded qualities are all the so-called proper objects of the senses. Hence, a classical observer must sort out what is objective from what is subjective in the experience of observation just to find the ground for the use of the language of classical mathematics in the first place.

The classical analysis, in other words, tries to understand the process of sensation from outside in, exactly as it does any other physical process. All the senses are modeled as a process of contact. (As Merleau-Ponty put it (1964a: 170), Descartes analyzed vision as if the blind were to "see with their hands.") Recall the passage from Leibniz about the machine that is supposed to perceive. There he tried to imagine, exactly as in the above analysis of vision, how to construct sensual experience out of body understood from outside in, with outside-in pieces, so that the experience would be a state of mechanical body.

Leibniz himself took the opposite view. To put it all too roughly, each mass point is real even if immaterial: a monad. From inside out each monad perceives the universe, though not all monads are souls. The "bare" ones are responsible for what we normally think of as body; they are passive from inside out, and thereby show up in the experience of souls as, for example, impenetrable. But for our purposes here the crucial aspect of Leibniz's theory is that, like Newton, he thought of the experience of the soul as entailing *a point of view*:

> And as the same city regarded from different sides appears entirely different, and is, as it were multiplied respectively, so, because of the infinite number

> of simple substances, there are a similar infinite number of universes which are, nevertheless, only the aspects of a single one as seen from the special point of view of each monad. (464)

Each monad is unique; the universe is referred to its point of view. So if it is a soul, a monad may well yearn to experience the universe from another's point of view, from inside, inside out. Classical theorists as far apart as Newton and Leibniz were both captivated by a point of view and the space of the world in view, even if Leibniz, unlike Newton, did not strip the point of view from the view.

All classical theorists faced the same dilemma: the reconciliation of an inside-out point of view with an outside-in body. Like Leibniz, one could take an inside-out point of view as fundamental, and attempt in its terms to account for an outside-in body. Or else, like Descartes and Newton, one could have both, essentially divided, and attempt to bridge the gap between them. Few followed Thomas Hobbes in taking an outside-in body as fundamental and attempting in its terms to account for an inside-out point of view.

For those like Newton, the inside-out orientation of a point of view gave rise to numerous problems, most notably the problem of deception. We very much wished to get beyond our points of view, to body in itself, though to do so we still had to work with our points of view: they must have some objectivity, for otherwise we could hardly ground our use of a mathematical language in the first place. For all the trust we placed in our senses in everyday life—again, as Hume advised, a little everyday experience is the only cure for skepticism—we had to be on our toes not to be lulled into believing our eyes. In the end, indeed, we wished to write down only the properties of body in itself, minus all the subjective, inside-out distortions of a point of view. We wished to understand how to take a point of view into account so as to eliminate it from the account.

Herein we cannot review the philosophical discussion of this classical analysis. Let it suffice to raise the crucial question: how do we know a point of view is of body? Perhaps the most ingenious way of resolving this question—due to Kant—is to argue that the very possibility of experience in which we can raise such a question already requires an answer, namely, to put it all too roughly again, were our points of view not of body, we would have no points of view at all. This answer requires, however, that we do not get beyond our points of view to body in itself; body is simply that of which we have points of view. Or alternatively, that of which we have points of view is all that we can mean by the term 'body.' With a twist of idealism that must remind us of Leibniz, this answer was hardly satisfying to the determined classical physicist, though philosophy itself was never again the same. (Recall that Newton had referred to body in itself, the "things without," as opposed to the motional pictures of them to which the soul has immediate access.) This answer also fits well with our general remarks about the origin of inquiry, as we will appreciate better in the final section of part two.

So, we wished to extend our everyday picture of body to its depths—and here I am thinking of an activity as simple as the everyday way of breaking a thing apart to inspect its "pieces"—even though, once at those depths, it was not clear how to reach the surface again, and any argument from the very possibility of experience seemed only to skim that surface. Still, the new physics worked so well during these times that it could not be challenged, and in the background lurked a benevolent God who could not have created us only to let us be subject to too much deception. We must be able to find a way to justify a physics that was otherwise so successful.

At least, following Newton, we knew that from God's point of view all our troubles were resolved. He was immediately present to, and so knew, what we had to presuppose: absolute space, time, and motion of body. (For Benedict De Spinoza space was an attribute of God, the only substance, and time no longer shared such substantial status.) Pierre Simon de LaPlace clearly imagined that, were he to have God's point of view, he would be able to infer with certainty any past or future state of the universe, but not because as God he would simply know these states; rather, once he had God's point of view, the universal, mechanical laws of motion would allow him to make such inferences (Capek, 122). Or again, with sufficient information about the universe at any one moment, he could apply the laws of motion as they were meant to be applied. This information must circumscribe the state of the whole universe at a moment: the state of the universe now.

In imagining that "the state of the universe now" represents an appropriate notion of simultaneity to employ in formulating the laws of physics, especially given a contact principle of connection between bodies, classical physicists displayed their profound spatialization of time. The state of the universe now is an event that excludes all other events, and each moment of time is correlated to such an event: it is as if time were simply a fourth dimension of space, and each state of the universe at a moment were always already in its place, some passed by and some yet to come. (Hence Spinoza eliminated the substantiality of time.) Here time entails an order of division analogous to space's order of division, which is expressed in the rule "no two bodies at the same place." Each state of the universe behaves in the same way: no two such events at the same moment. This notion of simultaneity assigns a unique event to each moment, or point of time, exactly as a unique mass is assigned to each point of space. Unlike the special theory, classical physics admits no problem, at least not in principle, in judging such simultaneity.

But let us also emphasize the irony of the classical analysis before finally turning to the special theory. How can we conceive of God's point of view?

We can use our own point of view: things in the space of the world in view stand apart at once. We simply extend this simultaneity as befits God. Hence, the irony is that the very ubiquity of our vision—think especially of color, again—is the paradigm of subjectivity on the classical analysis. (All along the classical analysis took the space of the world in view both to be indicative of the space of the world and to be deceptive, no doubt much too deceptive in the end. Here I am thinking of the position shared by Descartes and Berkeley, for example, that "distance, of itself and immediately, cannot be seen" (Berkeley, 1965: 285), and hence, as Merleau-Ponty put it in discussing Descartes (1964a: 172), "depth is a *third dimension* derived from the other two, height and width." This "distance" must be regarded as through the air—"a line directed endwise to the eye" (Berkeley, 1965: 285)—as opposed to along the ground, inasmuch as the latter *is* visible (Gibson). Captivation by points of view confounded us, especially when it came to the distinction between seeing and viewing.) Yet this is predictable, as we showed in part one: to succeed, experience must put us at a distance from the things in the world so that we can inquire about them, and this distance is all too easily converted into a division (or equivalently, into a construction) that is not so easily overcome.

It is no accident that so-called ancient thinkers, even when they divided up the world, tended somehow to introduce a way in which the divisions were overcome. These thinkers were only beginning to work with the quality of experience required for inquiry. They still made more of an order of separation without division than later thinkers would come to do. It is useful here to listen to Aristotle:

> Therefore not only can a thing come to be, incidentally, out of that which is not, but also all things come to be out of that which is, but is potentially, and is not actually. And this is the 'One' of Anaxagoras; for instead of 'all things were together'—and the 'Mixture' of Empedocles and Anaximander and the account given by Democritus—it is better to say 'all things were together potentially but not actually.' Therefore these thinkers seem to have had some notion of matter. (1947 (*Metaphysica*): 275)

This sense for wholeness, a residue of our lives before the dawn of modern inquiry, gradually faded away as we were drawn deeper and deeper into the quandaries of our own points of view.

III. Relativistic Analysis: Einstein

Everyday we tell the time. We do not begin our lives with the ability to do so, but sooner or later we acquire it. At the dawn of inquiry we learned to face the Paps of Jura at the proper moment to tell the beginning of the winter solstice. Now we learn to face a clock at the proper moment. Placed in front of a working clock, other things equal, we can simply look to see what time it is. The appropriate setting of the clock, not of the sun, must stand apart at once, in view.

But how do we know that a clock is working? We refer it to another working clock. The sun does not enter our minds, except perhaps to distinguish such large matters as night and day. We refer one clock to another; we synchronize them. And what time do they keep? Like other special standards in an equally special place, a certain clock must work to keep *the* time, though nowadays no one takes this too seriously. We have many different time zones around the world, more or less synchronized with the course of the sun, but we also have such phenomena as daylight saving time. We could set the special clock any way we wished, so long as the rest of the clocks were duly synchronized to it.

Let us suppose that we have the special clock in view: a certain setting of the clock stands apart at once. But also suppose we ask, "Doesn't light take time to travel from the clock to our eyes, and doesn't our nervous system take time to work with the interaction of light with our eyes, and so on?" Well, yes, when the clock reaches a certain setting we cannot be said to be, as Newton would put it (178), present to the thing itself. So, should we wonder what time it really is when the clock stands apart at once, with the setting in view?

Although in everyday life we certainly do not wonder at all, we cannot resist when we have in mind, as Newton did (17), that time may be defined without relation to anything external; we understand that in principle our view is in error, no matter how small or inconsequential the error may be. Newton supposed, let us say, that God tells the time defined without relation to anything external by being present to the thing itself. At the origin of inquiry we too felt present to the thing itself. Newton felt confined by "sensible measures," though he did not realize just how confined he was. Einstein knew better. We run into a fundamental problem even before we turn to wondering whether we can tell the time God tells.

Let us suppose that a certain clock keeps the time God tells, and that we need to synchronize other clocks with this clock. We place another clock side by side with the original clock in such a way that we can make a place away from which the two clocks stand apart at once, in view. Hence, we can safely disregard the time light takes to travel to our eyes; the duration in question applies to both clocks in exactly the same way. Being concerned only to rule out any special effects of light, we can also safely disregard the time our nervous system takes to work with the interaction of light with our eyes, and so on, though we must make a special note of our supposition that two side by side clocks *can* stand apart at once, in view. (We are obviously supposing as well that our views of clocks are far more precise that they actually are, but the principles of the cases above and below are not thereby distorted.) If we aim only to synchronize two side by side clocks, we can achieve our aim: they are synchronized if and only if the same settings stand apart at once, in view.

The aim, put more generally, is to judge simultaneity *at our place*. Although our place is neither so extended as to include the two clocks nor so unextended as to be a point, the velocity of light is so large that we can safely assume that the settings of the two clocks and our view of them constitute "a point event." Hans Reichenbach spoke of two "signals" or "beams of light" and "an observer who is able to record immediately only the simultaneity of their arrival to his place." (57–8). In the case of two side by side clocks, we are able to record immediately the simultaneity of their settings. Their settings stand apart at once, in view, presumably because the associated signals arrive simultaneously at our place. In the case of one clock, the setting and our view of it must constitute another point event: we immediately record the setting. We need not bother, so to speak, to get inside these point events; there we are not concerned with the special effects of light.

So, consider the aim of judging simultaneity *at two distant places*. Let us suppose here that by "two distant places" we mean any two places such that, if a clock is at each place, the clocks cannot stand apart at once, in view. We cannot make a place away from both clocks at once; we can face them only in turn. Hence, we can no longer appeal to their standing apart at once—or to the simultaneity of the arrival of the associated signals—if we wish to dispel all concern for the special effects of light. How do we *know* that any two such clocks are synchronized? Even if we had first synchronized the clocks side by side and then carried them to their present places, we could not have checked that they were still synchronized unless we were able to judge simultaneity at these places.

Suppose further that, having anticipated this problem, we had arranged a mirror side by side with one clock in such a way that we could make a place away from both the clock and the mirror at once, and the second clock would appear in the mirror as if it were standing side by side with the first clock; just so could both clocks stand apart at once, in view:

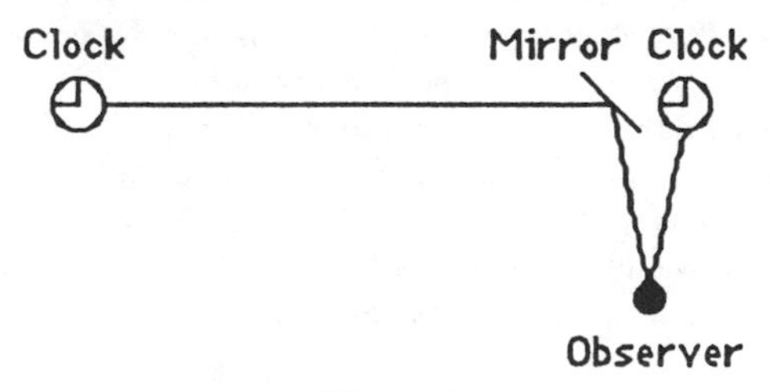

Figure 3

Do we know whether these clocks are synchronized? We simply take into account the necessary difference between the two, that of light travelling from the distant clock to the mirror. Again we do not have to bother with any special effects of light between the mirror and us. (Notice that, with a minor change, this technique will suffice for any case in which light must travel over different distances, including the one in which we can make a place away from both clocks at once but they are not quite side by side.) So, we know whether the two clocks are synchronized if we can take into account the necessary difference between the two.

Suppose even further that we had tried to anticipate this problem. The standard way of determining the duration in question is the same as the standard way of measuring the velocity between any two places. A clock is at each place, the clocks are synchronized, and the light takes just so much time to travel between the two places: the difference in time between the start at one clock and the finish at the other. But as we tried to carry out the standard procedure we ran into a snag. We had duplicated the very situation we were trying to anticipate. We had not as yet been able to tell whether two clocks at the very same places were synchronized. We needed the two synchronized clocks to measure the velocity of light between the two places, and we needed the velocity of light between the two places to synchronize the two clocks. We were caught—to combine the epithets of Reichenbach (59) and Einstein (23)—in a vicious, logical circle.

This special effect of light was first appreciated by Einstein. He asked (21): "On what basis do we have at our disposal the means of measuring time?" The basis must include the meaning of simultaneity or of simultaneous events; in synchronizing any two clocks, for example, we must understand what it is for them to have the same settings "simultaneously." Einstein had in mind the case of two lightning strokes at distant places when he stated the problem:

> We encounter the same difficulty with all physical statements in which the conception "simultaneous" plays a part. The concept does not exist for the physicist until he has the possibility of discovering whether or not it is fulfilled in an actual case. We thus require a definition of simultaneity such that this definition supplies us with the method by means of which, in the present case, we can decide by experiment whether or not both the lightning strokes occurred simultaneously. As long as this requirement is not satisfied, I allow myself to be deceived as a physicist (and of course the same applies if I am not a physicist), when I imagine that I am able to attach a meaning to the statement of simultaneity. (I would ask the reader not to proceed farther until he is fully convinced of this point.) (22)

The meaning of simultaneity is not as obvious as we thought it to be, even when we were thinking at our best, as in classical physics.

Let us return to the case in which we cannot make a place away from two distant clocks at once. Why not question the reality of this case? In everyday life at least, we may easily imagine that we can get a good view of these clocks without a turn of the head or a flick of the eyes or even a mirror. Here again we are captivated by making a place at a point of view; we tend not to resist the image of the two clocks standing apart at once in such a way that no movement or action is required to connect the two clocks to each other. We have "an idea of place separate from time": at a certain moment we have a view of places that are always already simultaneous. Einstein put it thus:

> On the basis of classical mechanics this four-dimensional continuum [of space and time] breaks up objectively into the one-dimensional time and into three-dimensional spatial sections, only the latter of which contain simultaneous events. This resolution is the same for all inertial systems. The simultaneity of two definite events with reference to one inertial system involves the simultaneity of these events in reference to all inertial systems. This is what is meant when we say that the time of classical mechanics is absolute. (149)

When we say in classical terms that two clocks at different places have the same settings simultaneously, therefore, the moment in question is not that of our view of the settings, but that of the two settings at those places, even though we cannot assume that all the events in question constitute a point event.

This meaning of simultaneity works very well in most circumstances. We do not notice, especially in everyday life, that these circumstances are circumscribed in a special way by the movement of light. Light moves so fast that we do not realize that, to take its movement into account in telling time, we reach the limits of the circumstances in which the independence of place and duration, of space and time, makes sense. If we are trying to measure a movement that is of the order of the velocity of sound, for example, then two clocks synchronized by reference to the order of movement of light—a simple radio signal, for example—can introduce only an error so small that we can safely neglect it, even entirely overlook it; the work of the light simply drops out of the final result. In everyday life, we tell how far away a lightning stroke is by counting the seconds from seeing the stroke to hearing the thunder. We never ponder the error introduced by the time light takes to travel from the stroke to us, exactly as we do not ponder the error introduced by the time light takes to travel from a clock to us. We feel present to the stroke itself.

We may try to get more accurate in this case by synchronizing two clocks, one at the stroke's place, and one at our place. Before we succeed, however, we will run into a special effect of light. With respect to the order of movement of sound we understand simultaneity by reference to the order of movement of light. But to what do we refer the order of movement of light? Yes, here we have no ready order of movement to which to appeal, save that of light itself! Certainly the orders of movement associated with our senses offer us no hope beyond light. Unlike Newton, Einstein understood what this confinement means. We cannot avoid the vicious, logical circle: we must appeal to the order of movement of light—*a movement*—to understand simultaneity, and we must appeal to a definition of simultaneity to understand the order of movement of light.

In classical physics we are—to put it in the terms of posture—swept off our feet into the space of the world in view, no longer in touch with the ground; we have a sense, not of a distance to be moved through, but rather of a distance separate from time and, in turn, of a simultaneity separate from movement. But actually to understand simultaneity we must appeal to an order of movement. The vicious, logical circle can therefore be said to require—again to put it in the terms of posture—the exchange of points of view for *standpoints*. From a standpoint a connection between events is always the result of an operation. In the special theory this operation is simply that of the movement of light itself.

To appreciate this conclusion, let us first consider how Einstein defined simultaneity in the case of strokes of lightning at two distant places. We certainly cannot be content with just any way in which the light from each stroke arrives at our place. So, before the lightning strikes at the two distant places, we arrange two mirrors at exactly half the distance between the places in such a way that, if we face the mirrors, the places stand apart at once, in view; "an observer placed at the midpoint" is able "visually to observe both places at the same time" (22):

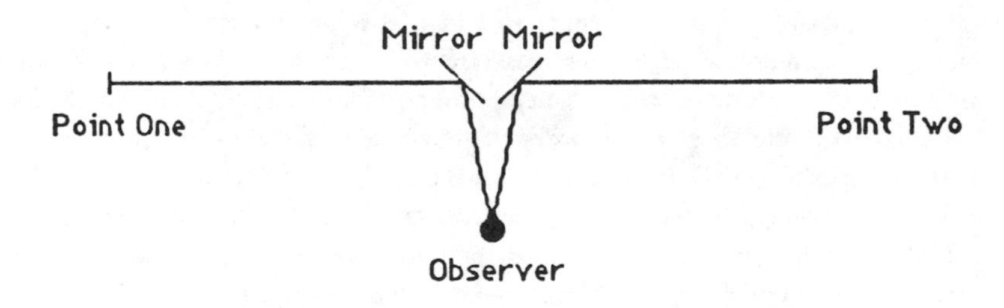

Figure 4

We may then state what we *mean* by the simultaneity of the two strokes: the strokes are simultaneous if and only if, for an observer who faces the mirrors, the strokes stand apart at once, in view. But again, why the stress on meaning? Isn't this merely a method for discovering simultaneity, rather than a definition of simultaneity? Can't we just hear someone who defends his or her purported discovery that the two strokes are simultaneous by saying, "But I was placed exactly halfway between them?"

Let us look more closely at the definition. Should we be reminded in any way of the first special problem about synchronizing two distant clocks? Well, certainly in this case the light from each stroke must move over the same distance. But mustn't we consider another aspect of the movement of light, that of its direction of movement? Do we *know* that the velocity of light is the same as it moves from each stroke to the midpoint? Well, no, we don't know, and if we try to determine either velocity, we will run in the same vicious, logical circle. Again, without a definition of simultaneity we cannot perform the required measurements.

Einstein did not regard this problem as resolvable by assumption or experiment: first we *define* simultaneity in terms of the two, equidistant mirrors and then we can tell whether two strokes of lightning are simultaneous. "That light takes the same time to traverse the path from one stroke to the midpoint as from the other stroke to the midpoint," he said (23), "is in reality *neither a supposition nor a hypothesis* about the physical nature of light, but *a stipulation* which I can make of my own free will in order to arrive at a definition of simultaneity." Why? Because "in reality the stipulation assumes absolutely nothing about light" (23). We cannot, not even in principle, put it to the test without already understanding what simultaneity is, or more generally, without already having "at our disposal the means of measuring time," yet it is precisely to have these means at our disposal that we make the stipulation. Hence, as Reichenbach also noted (60), we need not have chosen the particular stipulation that light travels at the same velocity in both directions. Why did Einstein choose it?

When Einstein required that an observer visually observe the two distant places at the same time, the moment in question must be the moment of the observer's view of the two places. Or again, the definition of simultaneity at distant places must be based on our ability to judge simultaneity at our place. We ask, "Does the light from the two strokes of lightning arrive at our place at the same time, simultaneously?" So, it may very well occur to us to try to measure the velocity of light on the same basis, even before we define simultaneity at distant places. We must then send light away from our place to be reflected back to our place so that we can note the time light takes for the round trip by referring to *one* clock at our place. We may also vary the direction in which we send the light so that we have a basis for comparing the velocity of light in different directions, even though this velocity must always apply to *a round trip* in each direction. The famous Michelson-Morley experiment involved just such a technique, with the aim of discovering a difference in velocity for round trips in different directions. Its result was negative. Light has always been measured to take the same time for round trips of equal distance, regardless of their direction.

Why did Michelson-Morley aim to discover a difference? Recall again that according to classical physics the movement of light must ultimately be referred to a body, the ether: light moves through the ether, we may imagine, as waves move through the water, or as sound moves through the air. Hence, should the ether be on the move itself, the effective velocity of light must vary in different directions depending on which way it is moving through the ether, though its velocity with respect to the ether remains the same, its velocity of propagation. Or alternatively, even if the ether is not on the move itself, the velocity of light will be measured to be different in different directions depending on which way an observer is moving through the ether; for example, should one be moving toward the light, its measured velocity must be greater than its velocity of propagation, whereas should one be moving away from the light, its measured velocity must be smaller than its velocity of propagation. Hence, this entire order of description is grounded on an ability to measure our velocity relative to the ether.

We began the Michelson-Morley experiment at some place on the Earth, let us say, by sending light away from us in perpendicular directions—one with the movement of the Earth—to mirrors attached to the ends of two identical metal bars:

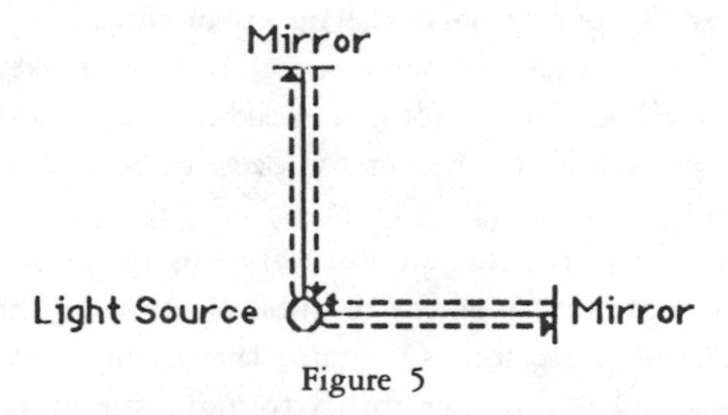

Figure 5

The movement of the Earth through the ether should have required a difference in the time light took to make both round trips. "The belated arrival of a ray," said Reichenbach (55), "could be proved by means of interference," in the form of a certain fringe shift that did not in fact occur. The velocity of light for both round trips was measured to be the same, regardless of the motion of the Earth through the ether. Or again, we could not measure our movement relative to the ether: it simply dropped out of the final result. Hence, the Michelson-Morley experiment failed to detect the ether.

Einstein saw that this result removed the only legitimate basis for believing in the ether. (Others had continued to believe in the ether even though they had to suppose that the effects of the ether cancel each other in such a way that the ether "cannot be demonstrated"; hence the original derivation of the Lorentz transformations (Reichenbach, 55).) Whenever we measure the velocity of light for round trips by judging simultaneity at our place, we fail to detect the ether, or equivalently, we do not determine our velocity relative to the ether. If we were able to judge simultaneity at distant places, we would also be in a position to determine a difference between the velocity of light for round trips and the velocity of light for one-way trips, and in turn to determine our velocity relative to the ether. But to judge simultaneity at distant places, we must stipulate the velocity of light for one-way trips, and in turn our velocity relative to the ether.

If, making a place at the midpoint between two distant places, we send light to each place and back, we will measure the round trips to take the same time. We will have neither a clue to any difference in the velocity of light in the two directions in question, nor a clue to our velocity relative to the ether. We will not be in a position to decide where to place the mirrors in our definition of simultaneity. We can *choose* a place only if we stipulate a velocity of light in each of the two directions in question, that is, only if we stipulate our velocity relative to the ether in such a way that, were we able, we would measure light to be moving at those velocities in those directions. Hence, our movement relative to the ether is a simple matter of stipulation, not of fact; it has nothing to do, as Einstein would put it, with the physical nature of light. The order of movement of light cannot have a preferred reference body.

Another way to appreciate Einstein's definition of simultaneity is to consider the motion either of the observer or of the entire situation. Let us suppose that two clocks are placed at the front and the back of a railway car moving along a track on an embankment. The mirrors are arranged at the midpoint of the car in such a way that an observer "placed at the midpoint" is able "visually to observe" both the front and the back of the car "at the same time": the clocks are synchronized if and only if the same settings stand apart at once, in view. Notice that an observer need not be at the midpoint before or after catching this view; that is, one may well be moving through the car in either direction so long as one arrives at the midpoint in the nick of time. Einstein's stipulation that light moves at the same velocity in both directions is independent of the motion of the observer or of the entire situation, the car, the Earth, or whatever.

The independence at issue here is crucial too: it is the basis for the celebrated relativity of simultaneity. Although the velocity of light is independent of the motion of a reference body, simultaneity is not. Let us suppose, as did Einstein as well (26), that two places on the embankment correlate to the front and the back of the car at the moment that lightning strikes at those places:

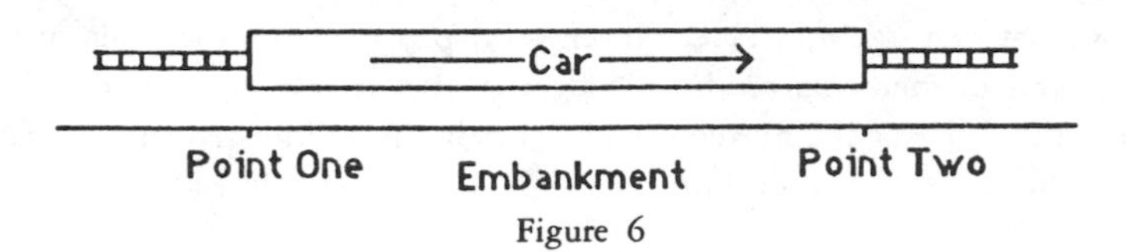

Figure 6

Again the definition of simultaneity may be employed by an observer at the midpoint of the car, but one sees the flash from the front of the car first, and the flash from the back of the car second: the flashes are not simultaneous. The stipulation is still at work, however, as light moves in each direction at the same velocity.

On a classical definition of simultaneity, no matter how far light moves in a second relative to the moving car, it moves the same distance relative to the embankment "*in each second* of time" (Einstein, 27). But on our definition of simultaneity no duration of seconds is independent of the motion of a reference body. Or again, we no longer identify a preferred reference body with respect to which light moves at its natural velocity, as we tried to do with the ether. Or still again, we no longer identify a preferred reference frame with respect to which light moves at its natural velocity, as we tried to do with the space pervaded by the ether, the space defined without relation to anything external.

Light always moves uniformly, at the same velocity in all directions, with respect to reference bodies in uniform motion themselves (acceleration is a matter for the general theory, not the special theory). An inertial frame of reference must also be defined in these terms: simultaneity in that frame must be referred to a reference body in uniform motion, with respect to which light itself always moves uniformly. Hence, all inertial frames of reference are equivalent. (Again we derive the Lorentz transformations.) Only on this basis do we make physical inquiry possible by defining how to measure time.

Still another way to appreciate Einstein's definition of simultaneity is to realize how it entails the relativity, not only of simultaneity, but of distance as well. The key is that, in measuring a distance between two places, we must consider the two places *simultaneously*. Standing on the embankment, let us suppose, we try to measure the distance from back to front of the moving car. If we allow time to elapse between marking the back and the front, the car will be measured to be bigger, even quite a bit bigger, than it would be were we to mark the back and the front at the same time; in the elapsed time the front will move along the embankment as far as the velocity of the train takes it. No, we must mark back and front at the same time. But as soon as this condition enters the picture, so too does the definition of simultaneity. If simultaneity is relative to a reference body, so too is a certain distance; we measure the latter in terms of the former.

Whereas in classical physics we need not appeal to an order of movement if we refer to places and moments, now we must appeal to the order of movement of light. Or again, whereas in classical physics both intervals of places and intervals of moments are independent of the motion of a reference body, now it is the order of the movement of light that is so independent. Whatever events light *connects* in a certain way relative to one reference body, it also connects in the *same* way relative to any other reference body. Now it is a *space-time interval*, a relationship *between events*, that is invariant. (Notice that different stipulations for light's one-way trips would produce the same space-time intervals; otherwise no stipulation would be necessary.) The order of movement of light is, as Reichenbach put it (67), "the ordering net of physics." We must stress *net* because of the interdependence of space and time, which constitutes a four-dimensional continuum essentially different from that of classical physics:

> The sum total of events which are simultaneous with a selected event exist, it is true, in relation to a particular inertial system, but no longer independently of the choice of the inertial system. The four-dimensional continuum is now no longer resolvable objectively into sections, all of which contain simultaneous events; "now" loses for the spatially extended world its objective meaning. It is because of this that space and time must be regarded as a four-dimensional continuum that is objectively unresolvable. (Einstein, 149)

The relationships *between* simultaneous events and the views that determine them, however, are just as invariant as any other relationships *between* events connected by the movement of light in a four-dimensional continuum.

Einstein's definition of simultaneity at distant places is founded on a certain event: an observer has a view in which these places stand apart at once, in a space separate from time. Nevertheless, the observer does not actually make a place in such a space, for the observer's place must be the destination of light from distant places, and light moves through space-time. Hence—to put it in the terms of posture—the observer cannot be said to make a place at a point of view, to be swept off his or her feet, no longer in touch with the ground. As we concluded in part one, only if we make a place on the ground, at a standpoint, are space and time inseparable with respect to that place. We have the *space* of the world from a point of view, but the *space-time* of the world from a standpoint. Or again, the observer cannot be said to make a place at a point of view in space-time, for no points of view exist in space-time: *in space-time views have no objectivity*.

Accordingly, the special theory undermines the positions of both Aristotle and Newton. Light cannot be "as it were the proper color of what is transparent"; the air that is transparent fills the space of the world from a point of view, not the space-time of the world from a standpoint. The problem of transparency—that the near and the far side of a transparency exist "at the same present moment"—cannot be resolved in terms of light alone. The same must be said about the problem of space defined without relation to anything external. The movement of light, or alternatively, a reference frame for the movement of light is always referred to a place in space-time, *to a standpoint*, from which we know absolutely nothing about other events "at the same present moment." As Einstein put it in the last quote above, "now" loses for the spatially extended world its objective meaning. Here Einstein was more thoroughly classical than Newton was: whereas for Newton the space of the world in view could still be objective to the extent that it constituted the paradigm of the absolute space in which light actually moved, for Einstein the space of the world in view could not be objective at all.

The order of movement of light circumscribes domains of space-time in which we can attach meaning to a claim that two events are simultaneous. Given a standpoint here and now the associated domain includes all those events which the order of movement of light (or of any movement slower than that of light) connects to the standpoint. This domain is the light cone of the standpoint, and the standpoint is regarded as a point event at the center of a frame of reference, though again the point event actually includes more than the standpoint.

From any one of our standpoints certain events fall outside of the net of events within the light cone of that standpoint. These events cannot be casually connected to us at that time and place. Here and now we cannot know anything at all about these events. They are, so to speak, not in our time; they are *space-like* separated from us, absolutely elsewhere with respect to our standpoint. Events within, or on the boundary of, the light cone of our standpoint are, to continue a manner of speaking, in our time; they are *time-like* separated from us, absolutely past with respect to our standpoint.

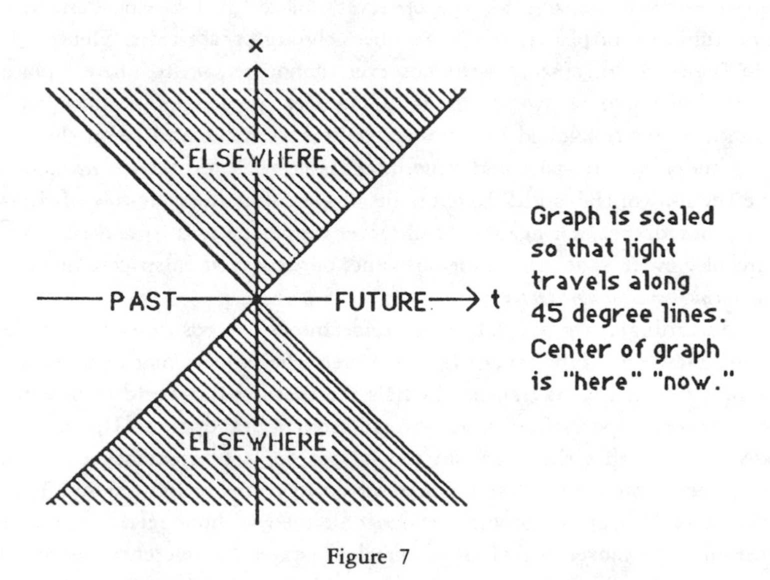

Figure 7

We say "absolutely" elsewhere or past because space-time intervals are invariant.

So, from a standpoint we do not order *the* past but only *our* past, those temporally prior events the news of which light can bring us here and now, the moment of our standpoint. Or alternatively, we can order only time-like separated events. It is important to understand that, given our stipulation about the velocity of light, a movement faster than light—a movement which could have brought us news of events space-like separated from us—cannot exist. This restriction is part of the *meaning* of simultaneity. Recall our earlier remarks about using light to determine the velocity of sound. Were a movement to be faster than light, we could use it to *determine* the velocity of light, in any *one* direction we wished to consider, by referring to *that* movement in order to stipulate what simultaneity at distant places means. (The movement would have to be able to carry a signal, of course, for otherwise we could not use it to synchronize two clocks.) By referring to a movement faster than light, we could convert our stipulation about the velocity of light into an hypothesis to be put to the test (and no doubt rejected were we really able to refer to a faster movement). We could separate the problems of measuring time and of measuring light; light would no longer be the ordering net of physics. But these problems are not separable. We must make the stipulation about the velocity of light, thereby connecting what we *mean* by "time" and "the movement of light."

Whenever the movement of light is measured, the result is a certain velocity, C (in a vacuum). We then stipulate that light always moves at C. Anything that moves slower than C cannot be light, cannot even become light, say, by adding energy to increase its velocity. As its velocity increases it *must* become harder and harder to move; its effective mass approaches infinity as its velocity approaches C. All movements with a velocity less than C can be brought to rest with respect to a frame of reference that moves with that velocity, and in that frame its effective mass becomes its rest mass. But the movement of light cannot be brought to rest, as it is stipulated to move at C in *all* frames. Light cannot have a rest mass, and anything that does have a rest mass cannot be light; it cannot move at C, nor come to move at C. This restriction, once more, is part of what we *mean* by "the order of movement of light," what Einstein called "an irreducible element of physical description" (150).

In order to measure time, we must displace Newtonian body from its privileged position as an irreducible element of physical description. Henceforth light, or rather, a field of light, which cannot be regarded as a state of Newtonian body, takes the privileged position. The field is an order *of* movement or motion, not *in* motion—or not of *that in motion*—though it must still be said to fill space-time. (Only on the general, not the special, theory of relativity does the order of movement of light become *the* irreducible element of physical description: space-time as opposed to what fills space-time "has no separate existence," as Einstein put it (155).) Or again, there is no ether. We have penetrated inside Newtonian body, only to find no such body at all: the field of light exists "in 'empty space' in the absence of ponderable matter" (Einstein, 145). Let us now turn to a discussion of this result.

IV. Relativistic Analysis: Implications

A rock has a rest mass. If it is at rest, a force has to be applied to move it, and if it is moving, a force has to be applied to bring it to rest, or even just to slow it down or speed it up. It has the normal inertial properties of Newtonian body.

Let us suppose we are investigating a rock that we have in view. We touch it; it feels solid, as if it were one body. But with a hammer we manage to break it into lesser rocks. We realize that its original rest mass was constituted by these lesser rocks. If we continue to break it into even lesser rocks, we will eventually realize that its original rest mass was constituted not only by the rest mass of other bodies but also by the motion of those bodies. That is, as we approach the original rock from outside in, we eventually encounter the phenomenon of inner movement, which contributes in a relatively hidden way to the original rest mass. At some fundamental level of organization of the rock we picture quite tiny bodies on the move, each possessing, as a result of its movement, an effective mass somewhat larger than its rest mass.

So far, then, we have not exceeded the limits of Newtonian body. We still picture bodies that behave as the original rock would do were it that tiny. We may very well think of them as bouncing off each other as the particles of the air in a tire bounce off each other and the walls of the tire in the process of keeping the tire inflated: from outside, the rock and the tire behave as if they were filled with body.

Ultimately we picture lesser bodies that we cannot break. However these bodies contribute to the original rock, they cannot be said to do so by their inner movement. At the limit of inner movement in Newtonian physics, indeed, we encounter what Einstein would call "an irreducible element of physical description," though here the irreducibility is not Einsteinian: a simple body taking up space and perhaps on the move in space as well. It is such a body, no doubt, that served as the classical paradigm of solidity.

Solid bodies at the limit of inner movement were not always taken to be rigid—some were taken to be elastic—though the notion of rigidity placed a close second to that of solidity in the classical paradigm. Why? Recall that in the Michelson-Morley experiment light was directed in two perpendicular directions—one with the movement of the Earth through the ether—so that it could be reflected back to the origin by two mirrors placed at an equal distance in each direction from the origin. Let us suppose, as did Reichenbach, that these mirrors were attached to two metal bars that spanned the two equal distances in question. If these bars are rigid, then light ought to take different times for the two round trips. Since no such difference was detected, Hendrik Lorentz tried to salvage the ether by supposing that one path of light must not be the same as the other in just the way required to cancel the effects of the different velocities of light, or equivalently, by supposing that the bars were not rigid because, as Reichenbach put it (55), "ether exerts shortening forces upon moving bodies in such a manner that the differences in the velocity of light connected with motion cannot be demonstrated." Even before Einstein's breakthrough, we could tell that the classical paradigm was in trouble, for it no longer could support a notion of rigidity. A similar insight must also occur once one understands both that communication across a rigid body would have to work instantaneously and that no communication works faster than the speed of light.

To step back a moment, we know that the mathematical treatment of a simple classical body depends upon the notion of a mass point. It is as if a body at the limit of inner movement were a field of mass points. (Only a perfectly spherical body acts as if its mass were concentrated at its center.) A field of mass points? Suppose that we try to take this picture more seriously. What would allow us inside Newtonian body? Whatever it is, it must not be "ponderable matter," as all the lesser bodies we have considered so far have been. Or again, it must possess no inside. Without an inside, however, how can it constitute body? Given this dilemma it is obvious why, until the dawn of the special theory, no one could give up on Newtonian body.

No inside? What is the function of an inside, anyway? Isn't it to allow a body to fill space while it is at rest? What about that which is *never* at rest? How does it fill space? Clearly, it can do so only by *pure* outer movement or displacement through space, together with the capacity, as such, to do work. Instead of taking body to be a field of mass points, suppose we take it to be a field of light points, that is, an order of a field of light. These light points, to continue a manner of speaking, would have no rest mass, no possible inner movement, merely outer movement. Einstein realized that the previous limit of inner movement could be crossed; he could imagine the mass of a body at that limit to be constituted by the movement of light—technically, electromagnetic radiation in general—exactly as we have imagined up to this point in Newtonian terms. (Think again of rigidity and the problem of instantaneous communication across a rigid body, but now fill the body with light instead.) Imagined to be in an appropriate box, the movement of light can simulate any rest mass from a perspective outside the box.

Unlike Einstein, we do not have to use our imagination to illustrate the equivalence of mass and energy that comes along with a deeper understanding of the two. We have discovered that even the most stable elementary bodies, the candidates for the fundamental Newtonian bodies, all have anti-particles (in some cases themselves). When a particle and its anti-particle meet, they may be annihilated in such a way that merely energy remains. On the basis of this result and others like it Bohm drew these conclusions:

> It is already clear that the "creation" of a particle [that is, an elementary particle] should correspond to setting up some characteristic relatively invariant kind of movement in a level below that of the elementary particles, with the aid of the necessary quantity of energy, and its "annihilation" to the ending of this pattern of movement, with the liberation of a corresponding quantity of energy. What seems essential here is to set aside the notion of "elementary" particles as the permanent substance of matter, and to regard them as only *relatively* fixed kinds of entities, which come into being when certain kinds of movement take place, and pass out of being when these kinds of movement cease. (1965: 120)

At the new limit of inner movement, however, we still encounter an irreducible element of physical description: under all transformations, from all standpoints that move uniformly with respect to each other, the velocity of light never changes, constituting an order *of* movement at that velocity, whose superstructure is a net of events specified, among other ways, in terms of mass and energy. Through the famous equation '$E = mc^2$' we describe mass and energy as *relatively* invariant with respect to the field of light, the electromagnetic field. (Again, in the special theory the field of light fills space-time, but in the general theory space-time as opposed to the field "has no separate existence." The implications we draw above and below from the special theory are in no way undermined by this additional, relativistic interdependence.) But what does all this mean when we apply it to the original stipulation: I am body inside out?

From outside I am like a rock; I, too, have the normal inertial properties. As we begin to penetrate inside me, we will find that lesser bodies, themselves in motion, all contribute to my rest mass. Again, at each such level, what just appeared to be one body becomes many bodies in motion. At the new limit of inner movement, "below" the Newtonian level of organization of body, however, we do not speak of bodies in motion, but rather of an order of movement of light which itself is pure outer movement. Fundamentally, like the rock, I must be organized at the limit of inner movement: in some way I am an order of movement of light, a field of light, with a superstructure of events that constitute the various levels of organization of body we discovered as we penetrated deeper and deeper into my body. Each level is understood, we can say, in terms of the appropriate relative invariance of the level "below" it. When it comes to my nervous system we may consider the relative invariance of the order of movement of the "elementary" particles of that system, though ultimately we must consider the relative invariance of the order of movement of light itself.

Now let us imagine that we "move" from inside out, from the irreducible level of organization to those levels of organization "above" it (that is, to the levels at which Leibniz might have imagined walking were they bigger). In this way, we understand how the latter levels are "built," each being a level of relative invariance with respect to the level "below" it. But I am also not like the rock: the level of organization of "an observer" will also have to be "built." In classical physics, we proceeded as if the observer could be eliminated from the universe he or she observed, as if the laws of physics could be applied without reference to the observer who applied them. But in the special theory, as Bohm concluded, we may no longer proceed in this fashion:

> *The observer is part of the universe.* He does not stand outside of space and time, and the laws of physics, but rather he has at each moment a definite place in the total process of the universe, and must be related to this process by the same laws he is trying to study. . . . In Einstein's point of view, it is clear that any particular Minkowski diagram [of space-time] is a map corresponding to what will be observed in a system moving in a certain way and oriented in a certain direction. Therefore, this map already has some of the observer's perspective implicit in it. (1965: 177, 183)

Indeed, it must do so, for only by reference to this perspective is it possible to understand how to interpret the map. We must also be careful to note, notwithstanding the typical use of the term 'perspective' to refer to a point of view, that the observer's place must be a standpoint, not a point of view. A standpoint centers the light cone of a Minkowski diagram, as the point event at its center: $(x,t)=(0,0)$. As time goes by the standpoint usually traces a "world line" (or a "world tube," were we to think of the extension of the observer). For simplicity in what follows we will confine our discussion to a standpoint that centers the light cone of a Minkowski diagram.

The point event at such a standpoint, we already know, must include a view that determines simultaneity with respect to that standpoint. This view—or as Einstein might have put it, this "now" that loses for the spatially extended world its objective meaning—must be the upshot of a lot of work in the past, including the movement of the relevant light to our eyes and the resulting activity of our nervous system itself. The view is "built" in this way. Bohm spoke of "a *construction* . . . based on the abstraction of what is invariant in the relationship . . . of the outgoing movements and the incoming sensations" (1965: 203, 213). ("We have seen that in perception there is present an outgoing nervous impulse producing a movement, in response to which there is a coordinated incoming set of sensations" (213).) Whether or not Bohm regarded this whole process of movements, sensations, and abstraction as the work of body, he still claimed that the construction "is present to our awareness in a kind of 'inner show' " (203). Hence "our awareness" is not somehow an element of the construction itself. Although here Bohm cannot help but remind us of Newton (as if the "inner show" were put on by our brain for the benefit of our soul), he nevertheless applied the principles of the special theory to the observer as well as to the other aspects of the universe:

> That is to say, just as the relativistic "map," in the form of the Minkowski diagram . . . must contain something in it to represent the place, time, orientation, velocity, etc., of the observer, so the mental map that is created by each person must have a corresponding representation of that person's relationship to the environment. (1965: 192)

This corresponding representation is required not only by the special theory but also by theories in other disciplines. Bohm gathered evidence from a wide range of sources, principally from the work of Jean Piaget and—given our own work in part one this source should come as no surprise—the work of Gibson. In these sources, Bohm discovered principles of "construction" that coincide exactly with the principles of the special theory (1965: Appendix). Hence, Bohm felt confident in talking about the abstraction of relatively invariant features from an intersection of the orders of outgoing movements and incoming sensations. Such abstraction is, as Bohm put it (1980: 153), "evidently very generally applicable."

What Bohm did not do, however, was clearly to confine himself to the work of body. Do we need to go beyond the work of body to account for the "inner show?" Why not begin to utilize theories such as the special theory to reconsider the nature of body, with an eye toward what it is possible for a human body not only *to be*—for, at the very least, we must be subject to the same levels of organization as the rock if we are to take our place in the same universe, under the same laws—but also *to use*? What if the "inner show" exists only because a human body *works* at the level of organization of itself which on the special theory we refer to as the limit of inner movement? The alternative that we do not use all of what we are seems surprisingly narrow once it is stated. Perhaps somehow, inside out, we do work "below" the Newtonian limit of inner movement.

Before we can consider this suggestion, however, we must be more careful about the nature of the "mental map." As already noted, we should not use a term like 'perspective' to characterize the observer's "relationship to the environment." As far as light and vision alone are concerned, this relationship cannot be one of a point of view to a view, notwithstanding these remarks by Melvin Rubin and Gordon Walls in their work on the fundamentals of visual science:

> The "place" of a visual-spatial point is given in two coordinates, direction and distance, where both of these are laid off from "the self" (i.e., from the *visual ego*). Things are seen from a view point, which should be pronounced view *point*, since the visual ego behaves as if concentrated at a point. (359)

Things are *seen* from a standpoint, not a viewpoint. The problem of the visual ego and its behavior, or alternatively, the problem of the view that determines simultaneity—the "now" that loses its objective meaning—is not a function merely of light and vision, as we will discuss in more detail below. Put another way, the problem before us here is the nature of the "mental map" of the Hopi, who are on their feet in the visual world, aptly described by Gibson. Rubin and Walls conflated two terms that we must keep separate, 'visual self' and 'visual ego,' with the latter reserved for the space of the world in view. (The visual ego is *merely of* the world, whereas the visual self has a nose *in* the world, as Gibson might have put it.) So, how shall we understand the distance laid off from the visual self?

Let us consider the light that we get in the way of in order to see. The light moves along the boundary of the light cone of our standpoint. With respect to what is absolutely past and circumscribed by the light cone, the light constituting the boundary of the light cone is *of* the cone, but not *in* it. This light may well bring us news of various events in the cone, but what of news of events not in the cone, yet of it? That is, what of the light itself? In what way, if any, can we be said to see it?

Recall once again the example of seeing a stroke of lightning and counting until we hear it. Both the sound and the light in question carry information about the stroke, but the light moves more rapidly, indeed so much more rapidly that we act as if our seeing the stroke and the stroke were simultaneous. We obviously do not act this way about our hearing the stroke and the stroke; the sound in question *moves between* the stroke and us. Or again, we act as if the light in question were to fix the starting point of the movement of sound between the stroke and us, affording us an experience of the whole movement of the sound to us.

The light, in other words, brings us news of the upcoming arrival of the sound. The light gets to us first, warning us of the arrival of the sound. The sound itself cannot warn us of its own arrival; it brings us news of its arrival only by arriving. Were we somehow completely confined to an existence at the order of movement of sound, nothing, not even the sound itself, could bring us news about its upcoming arrival. Any truly limit movement must occupy a unique place in our experience: we can have no news of its upcoming arrival until it arrives itself, but then it has already arrived.

Or again, were we somehow completely confined to an existence at the order of movement of sound, we could not *measure* its movement to us, a one-way trip. To synchronize the two clocks required for this measurement we would need to appeal to the very movement of sound in question, thereby throwing us into a vicious, logical circle. We get out of this circle of sound only because we can appeal to the order of movement of light. But we do not get out of the circle of light; we are confined to an existence at the order of movement of light. Hence nothing, not even light itself, can bring us news of its upcoming arrival; it brings us news of its arrival only by arriving itself. The order of movement of light must occupy a unique place in our experience.

With no warning of the light that arrives at our place, we cannot resolve its movement in experience. To resolve a movement in experience we need information about that movement at more than one place or moment, as in the case of thunder: hence, the special role of the light's warning of the arrival of the sound in question. (Indeed, as we will discuss below, the uncertainty principle requires that we *cannot* have the required information at one place or moment; it makes no sense to speak of movement at one place or moment.) Our body is an instrument like any other; it cannot resolve a movement that, in principle, cannot be measured by an instrument. With respect to light, we can use our body only as we do the one clock in the Michelson-Morley experiment; that is, we can use it to resolve the movement of light only for round trips. What does this mean?

Let us aim a flashlight away from us toward a mirror so that its light will be reflected directly back to our eyes. When we turn the flashlight on, we simply see the reflection, though we still imagine the light from the flashlight to move out and back as if we were seeing it do so. Indeed, we may easily imagine that, were light sufficiently slow, or were our vision sufficiently acute, we could somehow catch this out-and-back movement. What is wrong with our imagination here? Well, notice that according to our imagination we may *know* when the light turns around, as if we were able to count as the light moves out and back. But this knowledge is precisely what Einstein showed we can never have: were our imagination realizable, we would not need to *define* when the light turns around; we could *discover* it. It makes no sense, Einstein might have said, to speak of the availability of such information. Light drops out of experience more completely than any other order of movement—indeed, as completely as possible for an order of movement to do.

The light of the flashlight brings us news of its round trip only as a whole, and were light sufficiently slow, or were our vision sufficiently acute, we could experience only the lapse of time between turning on the flashlight and seeing its reflection. In practice, of course, we experience no lapse at all, and it is in principle impossible to experience a lapse in the case of the typical one-way trip of light from an object to us. It is no wonder, therefore, that Aristotle believed that "the parts of media between a sensory organ and its object are not all affected at once—except in the case of Light, and also in the case of seeing" (1955: 447a). Inasmuch as we are confined to the circle of light, our experience may resolve only the events *in* the light cone of our standpoint, that is, the pasts of lesser movements than that of light itself.

Although we do not as yet have an explanation for the transparency of the space of the world from a viewpoint, in which objects stand apart at once, we do have one for the transparency of the space-time of the world from a standpoint. Even on our feet we cannot have a sense of the movement of light in space-time, though the separation of the events we do have a sense of is always a matter of an operation, as the Hopi would put it. (Whereas the visual self faces the past, the visual ego faces the present, breaking the present from the past, a phenomenon that defies explanation on the special theory, indeed, a phenomenon that for many centuries kept us from understanding light at all.) The contrast to be marked here is between events in the space-time of the world from a standpoint and objects—Newtonian bodies—in the space of the world from a viewpoint. As Bohm put it (1965: 148), "the analysis of the world into constituent objects has been replaced by its analysis in terms of events and processes." It is on the latter analysis that we understand the nature of the transparency of the visual world.

This much understood, however, we have not yet penetrated very far into the point event at the center of the Minkowski diagram. We have tried to understand why the "inner show" or "construction" does not include a sense of the movement of light from our environment to our eyes. What of the ensuing events between our eyes and the final "construction?" Consider these additional remarks by Rubin and Walls:

> An oculocentric "directional sign"—that which triggers perception in a particular direction—is the activation of a particular area-17 cell [a cell of the occipital cortex]. Each of these direction-giving cells "sees" *as if it* were looking through a minute window in the retina, or as if it were in the retina in the position of the receptor itself. (360)

We are asked to imagine, presumably, that the final "construction" does not occur until a certain cortical activity occurs. (The cortical activity no doubt extends beyond the activation of the area-17 cells.) Now we have the problem, not only of the lack of a sense of the movement of light through the distance laid off from the visual self, but also of the lack of a sense of the activity of the visual nervous system itself. All of this visual activity, of course, is regarded as inside the point event at the center of the Minkowski diagram.

What can we learn by thinking of this center instead as the activation of the appropriate cortical cells themselves? For the "as if" of Rubin and Walls we substitute another: it is as if the visual nervous system—perhaps the entire nervous system—were to *work* at the limit of inner movement in such a way that its working there must drop out of our experience as completely as the working of light up to our eyes must drop out of our experience. In the final two sections of part two, it should be noted, we will push the "as if" beyond even the limit of inner movement. Our speculation in the present section plays a supporting role to our fundamental speculation about the nervous system: it begins to call into question our seduction by outside-in images of the nervous system.

Recall our stipulation about being body inside out in terms of the special theory: at the limit of inner movement, at the "irreducible" level of organization of body, we *are* a certain field of light, a field that exists "in 'empty space' in the absence of ponderable matter." Rubin and Walls clearly did not take the nervous system to work at this level of organization; the relevant nervous system activity might as well be confined to a Newtonian limit of inner movement for which the activity requires a material medium—the ether of nervous system activity, namely, certain cells, or even certain "elementary" particles—and a privileged frame of reference in which the medium can be at rest. Hence, the question at issue here is this: can our nervous system be adequately understood in terms of "ponderable matter" alone?

Rubin and Walls could think of a particular cortical cell as if the cell itself were to see, but only because the movement of the nervous system from the eyes to the cell fails to show up in experience just as thoroughly as the movement of light from the environment to our eyes fails to show up in experience: the nervous system, so to speak, gets out of the way of itself in vision as if no movement at all were to occur between the eyes and the cortex, or better yet, as if the relevant movement were to take place at the limit of inner movement. Hence, on this analogy, we understand the transparency of the nervous system by going beyond the Newtonian limit of inner movement. From "below" this level of organization of body, the question of how body gets out of the way of itself in vision can no longer be posed: quite simply, inside out, no such body exists.

This analogy does not require that the work of the nervous system from the eyes to the cortex proceed at the velocity of light in a vacuum. Neither does the work of light from the environment to our eyes proceed at the velocity of light in a vacuum, though it comes close. The nervous system does not even come close. (This fact is often mentioned, even stressed, by neurobiologists such as David Hubel and F. H. C. Crick.) The force of the stipulation that we are body inside out, however, depends merely on the levels of organization of body: on the special theory, body must be organized in *some* way at the limit of inner movement, "below" whatever relatively invariant pattern of energy or mass constitutes the "elementary" particles of the nervous system. What is it like to be such a nervous system inside out, from "below" these particles? The stipulation leads us to speculate that it is like the transparency of space-time on our feet, with the visual self in control. (Notice that the speculation need not be confined to human beings; many creatures indeed may enjoy the transparency of space-time on their feet.) Inside Newtonian body we find an order of movement that always drops out of experience.

When neurobiologists say that vision is a trick played on us by our brains (Crick), they themselves must be careful not to be seduced by the various images of the brain that populate their work, especially the images in photographic enlargements (for example, electron micrographs). These images are all constructed from outside in, and depend as much on resolution in vision as do our everyday images of our body and the environment. Are they too tricks played on us by our brains? When we wonder how the brain works, we must be open to the possibility that, as Crick himself suggested, "our entire way of thinking about such problems may be incorrect." Our stipulation, together with the speculation it leads us to make, is a new way of thinking about a crucial aspect of how the brain works: *it gets out of the way of itself.*

Neurobiologists have not yet abandoned the mode of inquiry that led Newton to think in terms of a series of vibrations which cannot get out of the way of themselves, exactly like Bohm's "construction" cannot get out of the way of itself—hence the reference to an awareness that provides the required transparency. (Think here of one Newtonian body *bumping* into another such body. It is surprisingly hard to give up something like this image of the interactions of body, even when we know better.) If we wish to avoid the Newtonian duality here, we must insist that the required transparency also be the work of body. Yet this sort of work must still capture what Newton and Bohm had in mind; namely, the work must involve different levels of organization in such a way that, as a whole, it accounts for one level being *of* another. In terms of our stipulation, if one level is of another level, the first is "below" or inside the second. We speculate that from inside out, from "below" the Newtonian limit of inner movement, *transparency is the very nature of body*, though this transparency, we must always remember, is that of events in the space-time of the world from a standpoint, not of objects in the space of the world from a viewpoint.

Again, why do we still suppose that the most profound questions about nervous system activity will be answered in terms of an ether theory of that activity? Why not take seriously that *there is no ether of nervous system activity*, that this activity cannot be understood in terms of "ponderable matter" alone, or more specifically, that the nervous system *works* at the Einsteinian limit of inner movement? (Some light even gets directly to the cortex, exactly as some light—much more light, in this case—gets directly to our eyes. We are just beginning to realize how important the former case is to us: we are sensitive to subtle changes in the light that reaches the cortex directly. The speculation in question here simply takes us one step further; we wonder how the cortex itself is organized at the limit of inner movement.) This speculation naturally arises from our stipulation about being body inside out, and just as naturally provides an explanation for how body gets out of the way of itself in vision. Deep inside it is as if we were light, an order *of* motion, not *in* motion, or not of *that in motion*. Ultimately we are not Newtonian bodies in the space of the world from a viewpoint; *we are Einsteinian events*—very special ones as well—in the space-time of the world from a standpoint.

So far, of course, we are still telling the story of vision from outside in: we are following light itself outside in. We have not as yet penetrated "below" or inside light, to a level that is of it as it is of Newtonian body. But we do know that getting inside light will not be like getting inside Newtonian body, for light has no rest mass: from outside in, light has no inside. We also know that we have yet to understand how objects stand apart at once, in view, and break the present from the past. Let us now turn to the problem of understanding the visual ego and the associated transparency of the world in view, by considering the nature of body "below" the limit of inner movement.

V. Quantum Mechanical Analysis: Bohr and Bohm

On the special theory we no longer have the profound, classical divisions between all things in space and time. Although space and time are still separate from the "irreducible" field that occupies them, they are no longer separate from each other: hence, space-time. Moreover, each reference frame must be oriented in such a way that an observer, or rather, his or her standpoint takes its allotted place at the center of that frame. As Einstein said, once again, "'Now' loses for the spatially extended world its objective meaning." Along with this loss comes the special character of the field that occupies space-time: inside Newtonian body is an order of movement in terms of which we express, among other things, the equivalence of mass and energy. With respect to this order of movement the relationship *between* events takes on the invariant quality formerly invested in Newtonian body itself. Although events are separated in space-time, they are still connected as aspects of the superstructure of the "irreducible" field.

Measurement must work in this superstructure as well, but certain features of the old ways remain. Although a body may still be said to be here now, we may also need to refer to the whole contour of the "irreducible" field to understand what this means. Likewise, although a body may still be said to move from here now to there then, we may again need to refer to the whole contour of the "irreducible" field to understand what this means. "Being here now" and "moving from here now to there then" both remain as paradigms of the outcomes of measurements on the special theory, even though the relevant relativity enters into such measurements in every case. Certainly Einstein never welcomed any challenge to their significance: each event is localized in space-time, and its separability from other events is governed by the order of the limit movement (of signals).

To pose the challenge to localization and separability that Einstein recognized in quantum mechanics, let us now explore the details of one interesting measurement. We can easily recall how Einstein synchronized any number of clocks. He then said (24), "Under these conditions we understand by the 'time' of an event the reading (position of the hands) of that one of these clocks which is in the immediate vicinity (in space) of the event." Presumably both hands of the clock stand apart at once in an observer's view in such a way that no question of relativity arises here. No, the problem, if one exists at all, must concern the simple requirement that the "time-value" to be associated with the event in question is the reading *on the clock*. According to both the classical and the relativistic analyses, at the moment that an observer has a view of the reading on the clock the clock itself has moved along to *another* reading. If we think of the observer as claiming, at the moment of the view, to know what time it is here *now*, mustn't we worry about the discrepancy between these two readings, that is, about which "time-value" to associate with the event in question?

Well, *this* problem is not difficult to resolve. We can simply require that the event in question and the reading on the clock stand apart at once, in view. The places of the event in question and the reading on the clock must not be distant in such a way that any relativity enters into the situation. Hence, Einstein included the phrase "in the immediate vicinity (in space) of the event." We date the event in question as if we were synchronizing two side-by-side clocks: we synchronize the event, so to speak, with the clock at its side. The time delay between the reading on the clock and the observer's view of it simply drops out of consideration.

Yet we do attach a special significance to this time delay. Consider simply the situation of telling time: the reading on a clock (again, suitably synchronized with other clocks) in the immediate vicinity of an observer stands apart at once, in his or her view. When we are considering the special effects of light we assume that the time delay between the reading on the clock and the view can be neglected, but *we do not assume that no time delay exists*: neither light nor the nervous system works instantaneously. (No outside-in theory of vision can tolerate such a lack of duration. Here, obviously, the judgment of duration must be referred to a suitable standpoint in the immediate vicinity of the observer and the clock.) We may date the view of the reading and the reading at the same time, indeed, at the same time as any other event in their immediate vicinity will be dated by the side-by-side procedure, but we are not hereby claiming that they actually occur at the same time, as we *are* claiming with respect to the other events. We profess to *know* that the view of the reading and the reading do not occur at the same time.

We wish, then, to discover a procedure that takes this knowledge into account, or, in effect, a procedure for dating a view. A view is routinely taken to be one among other events in the so-called point event at the center of a Minkowski diagram, and outside this point event "now" loses its objective meaning. But, when we are not concerned about any special effects of light, we may wish to penetrate inside this point event to ask, as accurately as we can: "What time is it here *now*?" As soon as we ask this question *seriously* we notice that until we answer it we cannot give any *meaning* to the association of a "time-value" with a view, even though we are supposed to know, again, that such a "time-value" will be different from those we date by the side-by-side procedure in the immediate vicinity of that view.

We may assume that an observer knows what time the light leaves a clock of which he or she has a view: by the side-by-side procedure it leaves whenever the view indicates. We can then suppose that another clock, suitably synchronized with the first clock, is also suitably synchronized—somehow—with the observer's view of the first clock. Somehow? Well, suppose that some mechanism flashes a light when the view occurs, and that another observer has a view of the flash and the reading on the second clock. Then the time represented by this reading—again using the side-by-side procedure—can be compared to the time represented by the reading on the first clock. Have we come hereby to know the time delay in question?

Recall that, until we have, we cannot associate a "time-value" with a view. Then reconsider the mechanism that flashes a light when the view in question occurs. The working of any such mechanism will take time too, and we can ask about how much time. But we cannot answer until we know when the view in question occurs—we seek the time delay between the view and the flash—and this date is exactly what we are designing the mechanism to determine. We are running in a circle.

Think again about the mechanism here. To make such a mechanism we must have access to the view in question. We must be able to determine that the mechanism intersects properly with the view. What is proper here? The first event of the mechanism at work must occur at the same moment as the view, of course. How can we know this? Clearly, *if* we can we are in a position simply to apply the side-by-side procedure *directly* to the view, as if the view *itself* were to stand by a suitable clock. (Notice that, exactly as the determination of the time in the immediate vicinity of an event involves us in considering the *position* of the hands of a suitable clock, so too does the determination of the time of a view involve us in considering the *position* of the view, inasmuch as the mechanism must begin *there*.) Hence—at least in principle—the way out of the above circle is equivalent to a successful application of the side-by-side procedure to the view in question: the view must be localizable in space-time exactly as any other event is localizable in space-time.

Here we need to emphasize two points. First, we can now understand another reason why the observer is part of the universe: the observer must be open to investigation exactly as any other phenomenon is open to investigation if we are to make sense of assigning a "time-value" to his or her behavior.

Second, the localization of an observer's view of an event places it in the future of that event, at the end of the relevant causal chain. On the special theory, we must still approach the view from outside in, though the view can no longer be excluded from the universe. The determination of simultaneity with respect to a standpoint brings the view into the universe, only to have it analyzed as any other event is analyzed, from outside in. We can no longer pretend that a view standing side by side with a clock is *not in* the universe (because of, for example, its non-bodily character), but perhaps only God can get a good enough view of a view to date it.

The source of the basic problem here is this: following Einstein all along, we have referred to a view to date events while we are supposed to be on our feet in space-time. If we make a place on our feet, however, we do not really have a view, and if we have a view we do not really make a place on our feet. Or again, on our feet a view must be subjective—a trick played on us by our brains—at the very same time that it is indispensable for dating other events. Or still again, on the special theory we can give no explanation for the very transparency of a view in which two events stand side by side at once. (It is no wonder that, for Bohm, after the construction of an "inner show," the show must still be *present to* our awareness.) In the simple situation of telling time, a time delay exists from the standpoint in question, but not from the viewpoint in question; hence, on the special theory, the "now" from the viewpoint must lose its objective meaning from the standpoint.

But suppose that at another level of organization of body it is not possible to maintain the required, time-like separation of a view from the preceding events of the relevant causal chain. Time-like separation entails that the preceding events have definite properties that do not depend on the view to which they give rise. (In this respect, of course, the relativistic analysis is the same as the classical analysis.) So, alternatively, suppose that at another level of organization of body the properties of the preceding events cannot be specified apart from the view to which they give rise. Indeed, suppose as well that, given this other level of organization of body, we can explain views—and especially the transparency of views—in such a way that standpoints and viewpoints are no longer in an opposition that reduces the latter to subjectivity? Perhaps, in other words, "now" retains its objective meaning on this other level of organization of body. We will better understand this possibility after we consider the foundations of quantum mechanics. The remaining task of this section, then, is to prepare the ground for the next two sections in which we will explore this possibility.

The recent experiments based on J. S. Bell's Theorem have stimulated a lively discussion about the foundations of quantum mechanics. We need first to understand this discussion in its own terms. These terms will be represented here by the work of Bohm and Niels Bohr, both of whom did not entirely abandon the classical analysis. Hence, among other things, they concentrated on the role of artifactual instruments in quantum mechanics. We will also consider the work of the eyes regarded as instruments in themselves. Let us begin by briefly considering Bohm's account of Werner Heisenberg's thought experiment (1980: 128–134). Formulated prior to experimental tests based on Bell's Theorem, this account still anticipates their results in all the necessary ways.

In classical terms, we suppose that a particle is at rest in the target of a microscope. The microscope is designed to focus another particle deflected by the target particle so that the deflected particle enters a photosensitive emulsion and leaves what we observe as a track in the emulsion. We can illustrate the experiment as follows:

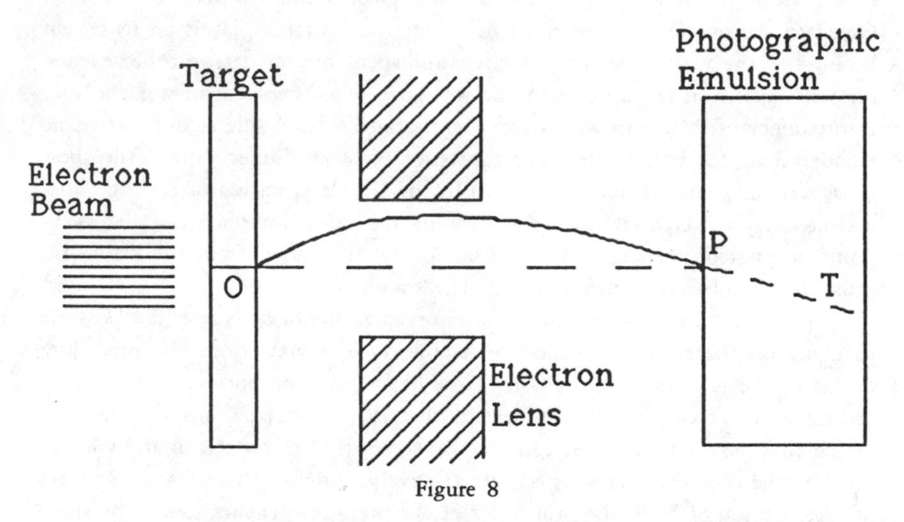

Figure 8

Given the track and an adequate description of how the microscope works—for example, how it affects the flight of a deflected particle—we make inferences about the properties of the target particle. On the special theory, the events that lead to the track in the emulsion are said to be within the light cone of the track; they are time-like separated in such a way that the past of the track can be uniquely determined. We come to know both the position of the target particle and the momentum imparted to it at the time of deflection.

Well, we come to know both if the classical and relativistic terms remain appropriate in all contexts. Notice especially the relationship between the experimental conditions and the final result. We imagine that the target particle in question possesses definite properties of position and momentum at the time of deflection, and moreover that its possession of these properties does not depend in any way on the experimental conditions themselves: the target particle and the experimental conditions are, as Bohm put it (133), "autonomously existent." Or again, as Berkeley would put it, the use of the microscope does not bring about "any manner of real alteration in the thing itself [that is, in the target particle]" (1979: 21). The description of the experimental conditions must drop out of the description of the final result.

Let us now consider the same experiment in a quantum mechanical context:

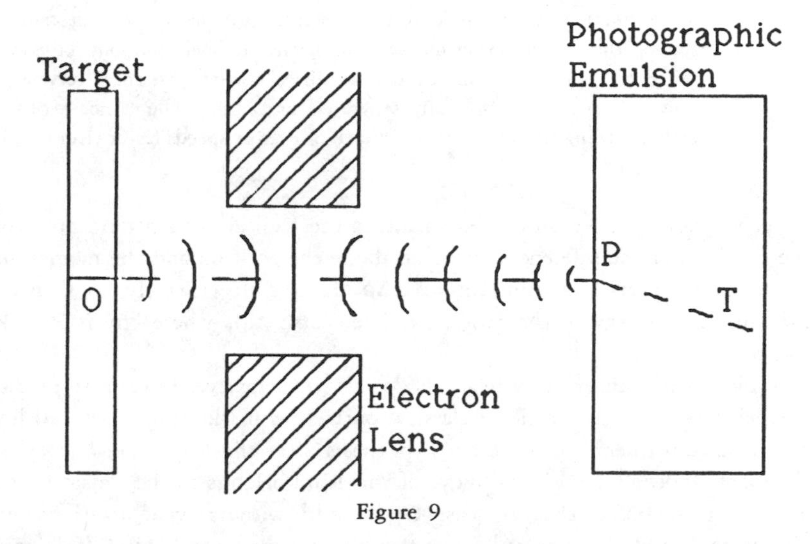

Figure 9

We suppose that the microscope is an electron microscope, and that the deflected particle is an electron. What must we say in quantum mechanical terms? On Bohm's account (131), Heisenberg "evidently brought in the four primarily significant features of the quantum theory":

1. He describes the link electron *both* as a wave (while it is passing from object O through the lens to the image P) *and* as a particle (when it arrives at the point P and then leaves a track T).

2. The transfer of momentum to the 'observed atom' at O has to be treated as discrete and indivisible. (The energy of an electron, for example, does not have a continuous range; the energy is expressed in whole multiples of the basic unit "E = hv," where 'h' is Planck's constant and 'v' is the frequency of the electron as a wave.)

3. Between O and P the most detailed possible description of the link electron is in terms of a wave function that determines only a statistical distribution of potentialities whose actualization depends on the experi-

mental conditions (e.g., the presence of sensitive atoms in the emulsion, which can reveal the electron). P can be predicted, but only with a certain probability, that of the potentiality that actualizes at P.

4. The actual results (the spot P, the track T, and the properties of the atom O) are correlated in the way indicated by Einstein, B. Podolsky, and N. Rosen, which cannot be explained in terms of the propagation of signals as chains of causal influence. In a more detailed mathematical treatment according to the quantum theory, the "wave function" of the "observed object" cannot be specified apart from a specification of the "wave function" of the "link electron," which in turn requires a description of the overall experimental conditions. (Abner Shimony (1984) coined the epithet "entangled potentialities" to refer to these correlations, which involve "spatially separated particles," though not necessarily anything like action at a distance or even at speeds faster than that of light.)

Together these four features of the quantum mechanical description imply that the precision of the inferences we make about the position and the momentum of the target particle is limited: $\Delta x \cdot \Delta \rho \geq h$ (or alternatively, in terms of the time and energy of the target particle, $\Delta t \cdot E \geq h$), where "h" is Planck's constant.

Now, these uncertainty relations do not in themselves force us to give up the belief that something like a classical particle is in the target. (Recall how we made adjustments to our ether hypothesis after the Michelson-Morley experiment. Indeed, the original move of this kind belongs to the classical physicists who supposed that at least God could witness what to them was undetectable.) Here we must believe that the quantum mechanical description of the particle is simply incomplete. Theorists such as Einstein, Podolsky, and Rosen wished to complete it with a hidden variables theory that could salvage the world of the special theory of relativity and its time-like causality, as opposed to the space-like correlations involved in the quantum theory. But now Bell's Theorem shows that no such hidden variables theory "can reproduce all of the statistical predictions by quantum mechanics." Although this result covers only "idealized situations," Bell's Theorem "can be extended to cover actual systems." John Clauser and Shimony asserted "with reasonable confidence that the experimental evidence to date"—and even more evidence exists today (Aspect *et al*)—supports the quantum theory over members of "the family of local, realistic theories" (1883–4). What exactly is this result?

Let us give the microscope experiment the form of "an idealized situation": for one track in the emulsion we can infer the position of the target particle with certainty, and for another track we can infer the momentum of the target particle with certainty. (Each track must be thought of, for reasons that will emerge below, as correlated to a different experimental arrangement, such as a different microscope aperture. For any one track, of course, the uncertainty relations exclude the possibility of making both of these inferences at once.) "How can this be," asks the family of local, realistic theories, "unless the target particle has this position *and* momentum all along, even though the quantum theory does not allow us to infer it? How can we suppose that making the track in the emulsion actually disturbs or acts on the target particle, especially at a distance?" The quantum mechanical description is then supposed to require supplementation by a local, realistic theory that (1) rejects action at a distance—though to be non-local a theory need not accept action at a distance—and (2) assigns definite properties to the target particle in virtue of our being able to predict with certainty that it has them without in any way disturbing it. The whole experimental situation is again analyzed into autonomously existent elements.

After presenting the evidence that this local, realistic theory is no longer tenable, Clauser and Shimony claimed that "Bohr's position remains as one of the few feasible options concerning the foundations of quantum mechanics" (1922). While accepting (1) Bohr did not accept (2). He understood that the mutual exclusion of the tracks on the basis of which we make the two inferences in question—one to the exact position, the other to the exact momentum—raises profound questions about assigning both properties to the target particle at once. Each measurement of the link electron, as he would have put it, affects "the very conditions which define the possible types of predictions regarding the future behavior of the system" (Bohr, 138). (Bohr had in mind the actual experimental arrangements required to measure the position or the momentum of a particle; whichever we choose to measure, an "uncontrollable" element in the measurement excludes the possibility of predicting the other.) Hence, the elements of the whole experimental situation are evidently not autonomous: that each measurement or experimental arrangement must exclude one track or the other deprives us of the opportunity to make certain predictions about the target particle, and in this sense alone, without action at a distance, we do disturb the target particle.

In general Bohr concluded that position and momentum—or alternatively, time and energy—are "complementary physical quantities" in such a way that any experimental procedure capable of providing an "unambiguous definition" of one such quantity excludes all experimental procedures capable of providing a definition of the other, and vice versa (139–40). This complementarity undermines (2) of the local, realistic theory:

> In a narrow sense, one can predict the value of a quantity only *when an experimental arrangement is chosen for determining the value of that quantity*. In a broad sense, one can predict the value of a quantity, *if it is possible to choose an experimental arrangement for determining it*. If the narrow sense is accepted, then the argument of Einstein, Podolsky, and Rosen does not go through, since the experimental arrangements for measuring the position and momentum of a particle are incompatible. (Clauser and Shimony, 1885)

Einstein, Podolsky, and Rosen had defined prediction as if a quantity and the experimental arrangement for determining its value were autonomous. Here, we can say, Bohr provided a more Einstein-like interpretation than did Einstein himself. The uncertainty relations circumscribe what we can *mean* by the terms 'position' and 'momentum': neither has any meaning beyond the possible experimental procedures appropriate to them.

Bohm put it as follows:

> Bohr gave a relatively thorough and consistent discussion of the whole situation, which made it clear that the four primary aspects of the quantum theory as described above are not compatible with any description in terms of precisely defined orbits that are "uncertain" to us. We have thus to do here with an entirely new situation in physics, in which the notion of a detailed orbit no longer has any meaning. Rather, one can perhaps say that the relationship between O and P through the "link" electron is similar to an indivisible and unanalyzable "quantum jump" between stationary states, rather than to the continuous though not precisely known movement of a particle across the space between O and P. (1980: 132)

To draw out the relevant aspects of Bohr's discussion Bohm worked with the following figure (132–4):

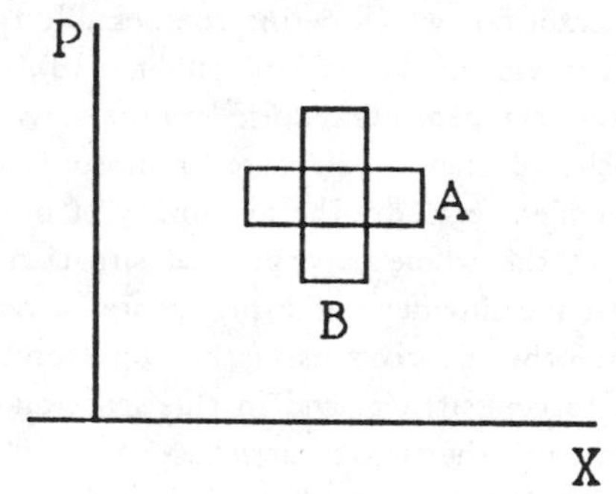

Figure 10

A and B are cells in the phase space of the observed object. Each has area *h*, and is associated with a different set of experimental conditions (for example, a different microscope aperture). Any such cell can be said to indicate the limits of applicability of the classical description under the associated experimental conditions: $\Delta x \cdot \Delta \rho \geq h$. In the classical context, the shapes of the cells have no significance because quantities of the order of *h* can be neglected. But in the quantum mechanical context the shapes of the cells remain relevant, as essential features of the description of the observed object.

The observed object cannot be properly described, Bohm concluded, except in connection with a description of the experimental conditions:

> This means that the description of the experimental conditions does not drop out as a mere intermediary link of inference, but remains inseparable from the description of what is called the observed object. The "quantum" context thus calls for a new kind of description that does not imply the separability of the "observed object" and "observing instrument." Instead, the form of the experimental conditions and the meaning of the experimental results have now to be *one whole*, in which analysis into autonomously existent elements is not relevant. (133, my italics)

Along with our tendency to regard the observed object as a classical particle comes another tendency to regard the relationship between the observing instrument and the observed object as a classical cause and effect relationship, perhaps even at a distance. Hereby we suppose that we can give unambiguous definitions to the observing instrument and the observed object in isolation from each other as autonomous existents. But precisely this lack of ambiguity is excluded in the quantum context: we *must draw* the line separating the observing instrument from the observed object. The necessity of drawing this line is, as Bohr put it (140), "the principal distinction between classical and quantum mechanical description of physical phenomena."

Bohr also believed that this dependence has "its roots in the indispensable use of classical concepts in the interpretation of all proper measurements, even though the classical theories do not suffice in accounting for the new types of regularities with which we are concerned in atomic physics"; we use quantum mechanics "to predict the results obtained by a given experimental arrangement described in totally classical terms" (140–1). What exactly does it mean to adopt the new limits to the meaning of such terms as 'position' and 'momentum?' According to Clauser and Shimony (1883, 1921), these limits undermine "the whole realistic viewpoint": "the term 'reality' can be used unambiguously in microphysics only when the experimental arrangement is specified." Notice nevertheless the curious lack of any specification regarding the person who arranges the experiment.

In the same spirit, Bohm did not extend "the whole" of the Heisenberg microscope experiment beyond the track in the emulsion. To discriminate between the observing instrument and the observed object, Bohm regarded the description of the track in the emulsion as the condition that resolves the relevant wave function. He spoke of an order of distinctions "relevated (and recorded) by our instruments," in this case by the emulsion (1980: 155). (Presumably, the track in the emulsion was imagined to exist, and to be correlated to the properties of the observed object, even without being observed itself. The question at issue here—surely representative of the question generally debated by quantum theorists—is precisely whether or not the observer of the track is as much an aspect of "the whole" as the track is.) With respect to the track in the emulsion Bohm was not concerned about quantum mechanical effects in the same way that he was concerned about them with respect to the interaction at the target. Witness the following remarks:

> "Elementary particles" are generally observed by means of tracks that they are supposed to make in detecting devices (e.g., photographic emulsions). Such a track is evidently to be regarded as no more than an aspect appearing in immediate perception. To describe it as the track of a "particle" is then to assume in addition that the primarily relevant order of movement is *similar* to that in the immediately perceived aspect. (154–5, my italics)

Despite indicating that the track in the emulsion cannot be attributed to a classical particle, Bohm did not resist a classical comparison between the immediately perceived aspect and the relevant order of movement in the emulsion, as if an analysis were to have established their autonomy in the way required to compare them to each other. Bohm also said that we "directly observe" the track in the emulsion (130). He must have meant to contrast this directness to the manner in which we observe the interaction at the target. Again, we must understand that with the track in the emulsion we return to a classical realm in which we do not directly observe anything but what is immediately perceived, the "inner show."

Bohr compared the problem of uncertainty to "the dependence on the reference system, in relativity theory, of all readings of scales and clocks" (141). In the special theory of relativity, we return to a classical realm when we look at clocks and scales in our immediate vicinity, and in the quantum theory we return to the classical realm when we look at emulsions: hereby we discriminate between the observing instrument and the observed object. Although Bohm did not agree with Bohr that the results of experiments could be described in totally classical terms—Bohm's own version of the lack of separability in the quantum context aimed more to found a new family of realistic theories than to abandon "the whole realistic viewpoint"—Bohm still did not extend "the whole" to the observer in the experiment. Let us now turn to an evaluation of this position.

VI. Quantum Mechanical Analysis: Implications

Taking our cue from Bohm, we start by noting what the track in the emulsion *is*: the relevant order of movement cannot be that of a classical particle. If we are looking at the emulsion, for example, mustn't we treat the properties of the particles of the emulsion in the target of our eyes as an order of quantum mechanical description analogous to that of the properties of the particle in the target of Heisenberg's microscope? The answer requires another thought experiment, following Heisenberg's very carefully: the particles of the emulsion in the original thought experiment are the observed object, an eye is the observing instrument, and we make inferences about the properties of the observed object. We assume that the particles of the original emulsion are initially at rest (relative to the eye), and that a light wave of known energy is directed at them in such a way that the wave is diffracted (through a vacuum) to the eye, which in turn focuses the wave in such a way that it leaves a track in the emulsion of this thought experiment, the retina of the eye. We can illustrate the experiment as follows:

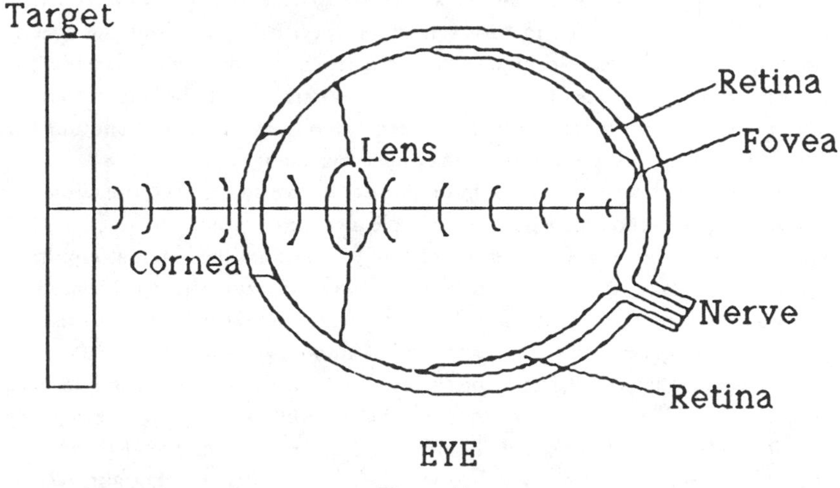

Figure 11

Given the above figure, we immediately realize the comparison to the original thought experiment: *the eye is a Heisenberg microscope.*

Indeed, the role of the retina as the emulsion in this additional thought experiment need not have been put forward as an assumption at all. As Bohm said of the original emulsion (1980: 131), the retina contains "sensitive atoms" able "to reveal" incident photons, though obviously not in the way that Newton imagined. Let us concentrate on the rods, which contain the photochemical rhodopsin:

> The absorption of quanta by the visual pigment of receptor cells leads to at least one chemical event that is indispensable for triggering a light sensation. In retinas that contain rhodopsin, when a quantum is absorbed a molecule of the latter shifts to prelumirhodopsin, i.e., from cis-form to trans-form. This first chemical event is followed by many others, but as yet there is no proof that the latter are indispensable to vision. . . . The absorption of a quantum

> of any wavelength must given rise to the same sequence of events in a given rod receptor, for the spectral efficiency curve of scotopic vision [the threshold of which is moonlight], after correction for light losses in the ocular media, coincides with the absorption curve of rhodopsin. (Baumgardt, 30)

A shift from the cis-form to the trans-form of rhodopsin constitutes, let us say, a track in the retina. Although this track is not as accessible as the track in the original emulsion, we still know that, were we to observe it, we could make inferences on that basis about the interaction at the target, namely, the original emulsion itself. To paraphrase Bohm, the wave function of the particles of the original emulsion cannot be specified apart from the wave function of the link photons, which in turn cannot be specified apart from the overall experimental conditions, including the track in the retina. By correlating the entry into the retina with the properties of the particles of the original emulsion, we are, again, discriminating between the observing instrument and the observed object in this additional thought experiment.

Accordingly, what keeps us from taking an eye to be the observing instrument in *both* thought experiments? We now know that the track in the retina is correlated to the properties of the particles of the original emulsion in exactly the same way as the track in the original emulsion is correlated to the properties of the target particle. But what *is* the track in the original emulsion—or perhaps more carefully, what can we *say* that the track is—but the properties that the particles of the original emulsion can be said to have on the basis of being "directly observed" by an eye? Hence, why not suppose that "the whole" of the original thought experiment can be extended into the eyes of its observer? We imagine—and this is the additional thought experiment, remember—another observer who uses the track in the first observer's eyes to make inferences about the properties of the particles of the original emulsion, and in turn about the properties of the target particle. Herein we do not make any assumptions about what happens in the vision of the first observer after the track in the retinas. We assume only that a track in the retina can be used exactly as a track in an emulsion would be used in any version of the original thought experiment.

Although in extending this experiment we are concerned with an order of error than can be neglected—and here I am thinking about the difference between what the first observer concludes on the basis of direct observation of the original emulsion, and what the second observer concludes on the basis of the track in the retinas of the first observer—we are also concerned with *another order of description*, the one that both Bohr and Bohm were willing to apply only to the experimental arrangement. Even if we do not extend "the whole" of the original thought experiment all the way to the retinas of the first observer, the quantum mechanical order of description is required by a crucial portion of the very process of vision that is supposed to be capable of description in totally classical terms. Stretching from *any* target into the retinas of an observer, at least up to "the first chemical event," the relevant order of movement can no longer be described as "the continuous though not precisely known movement" of classical particles. The relevant order is "a whole" that cannot be analyzed into "autonomously existent elements."

On Bohr's position, experimental procedures capable of providing unambiguous definitions of certain pairs of physical quantities—either position and momentum, or time and energy—are mutually exclusive in such a way that a procedure for position or time excludes a procedure for momentum or energy, respectively. We therefore reach a limit of classical description. In every experiment we must discriminate between the observing instrument and the observed object; the discrimination is no longer merely a given, but must be chosen, so that the description of the experimental result—still expressed in classical terms, according to Bohr—includes the discrimination and in turn the whole experimental situation. Our additional thought experiment extends Bohr's position, not only to observation of experiments in particular, but to observation in general as well.

We therefore reach a limit, not of descriptions on the classical analysis, but of the classical analysis itself. Even without extending "the whole" beyond our skin—and here I am thinking of the skin as the place of the first chemical event of perception, as in the case of the eyes—we can no longer extract our bodies from the descriptions of the world we observe. We begin to draw the line that Bohr made so much of with our very move to observe the world, even before we reach our artifactual instruments: only by reference to our bodies can we give unambiguous definitions of physical quantities. The human body is simply one among other instruments, quite a bit more sophisticated, but not essentially different in any way that allows it to be free from inclusion in the descriptions of the results of experiments it performs.

This conclusion, which involves no speculation at all, is by no means confined to human beings. It also applies, for example, to bees and frogs. Exactly as we stated at the beginning of part one, we cannot regard a visual stimulus to be *already* constituted as if the face of a bee or a frog were broken apart from it, somehow merely *waiting* to be stimulated. Although the visual nervous systems of bees and frogs are significantly different from ours, they are not different with respect to the nature of the first chemical event. Hence, a visual stimulus to the face of a bee, or frog, or a human being cannot be specified apart from that face. In all three cases, as we also stated at the beginning of part one, we can no more break away the visual stimulus from the face and the source of the stimulus than we can break away the face and the source of the stimulus from the stimulus: they constitute "one whole." Light is a stimulus only in the world of a face, or rather, in *the-world-of-a-face*, not merely in the world, broken apart from that face. Here we have *an order of co-making*.

But what of our looseness? Unlike the bee or the frog, we can be said to put our faces in the light that stimulates our faces: for us, another light must exist, not the light in the-world-of-a-human-face, but rather the light of or away-from a human face in the world, the light we can be said to bring to the world. For Descartes this light is an "inward" light, broken away from both the world and human faces; hence, in this light, our faces and the world must be "inward" as well. Or again, as Bohm would put it, this light "is present to our awareness in a kind of 'inner show'." For Plato, on the other hand, this light is neither an aspect of an inner show nor broken apart from the world and human faces. Here once again is the crucial passage from Plato:

> As soon, then, as an eye and something else whose structure is adjusted to the eye come within range and give birth to the whiteness together with its cognate perception—things that would never come into existence if either of the two had approached anything else—then it is that, as the vision from the eyes and the whiteness from the thing pass in the space between, the eye becomes filled with vision and now sees, and becomes, not vision, but a seeing eye; while *the other parent* of the colour is saturated with whiteness and becomes, *on its side*, not whiteness, but a white thing, be it stock or stone or whatever else may change to be so coloured. (1957: 47, my italics, as quoted in part one)

Indeed, if "the whole" were to extend beyond the skin, through the entire visual nervous system, Plato's theory of vision would actually be more appropriate than Descartes' or Bohm's would be: the light we can be said to bring to the world would be of *human-faces-in-the-world*, exactly as the light we cannot be said to bring to the world is in *the-world-of-human-faces*. In both cases we would have *an order of co-making*.

At this point in our analysis, therefore, we must try to decide whether or not the classical analysis of observation ever becomes relevant at all. Can we still draw a line to represent the classical separability of an event from its observation? In the additional thought experiment, for example, can we draw a line at the track in the eyes of the first observer, at the verge of nervous system activity, so that the first observer's relationship to this track is analogous to the relationship we once supposed to exist between the first observer and the track in the original emulsion, that is, a relationship analyzable in classical terms? But what keeps us from imagining, on the other hand, another version of the additional thought experiment in which the second observer does not intervene in the first observer's nervous system activity until after the first chemical event? Can't we suppose that the second chemical event marks the beginning of the relevant track? From outside in, after all, what can we *say* that the first chemical event *is* if we intervene in the first observer's nervous system only after the first chemical event? Technically speaking, we may well continue to express the physical situation in terms of wave functions even through the first chemical event: did a molecule of rhodopsin absorb "a quantum" or not? Then again, what keeps us from asking an analogous question about the second chemical event? It too must be open to *some* description at the quantum mechanical level of organization of body. Hereby we continue to express the physical situation in terms of wave functions as we follow, from outside-in, the series of events in vision.

To illustrate how to create further thought experiments, let us consider the entire visual pathway, sketched here from below:

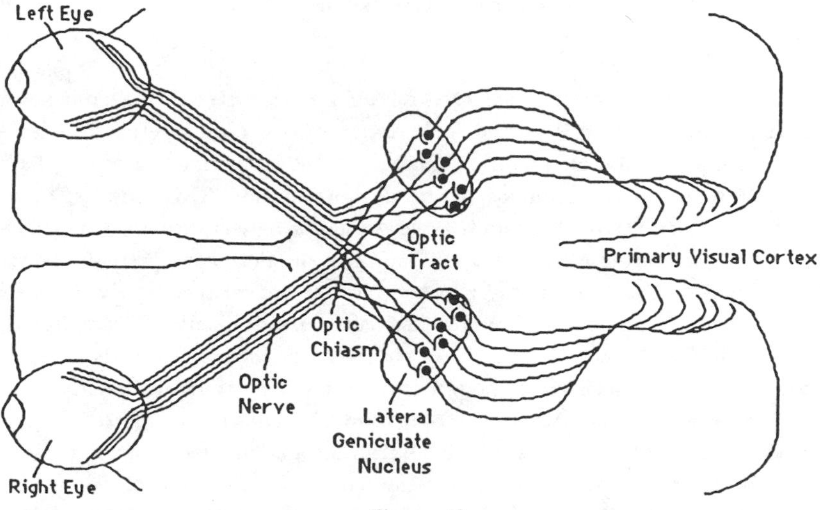

Figure 12

The most typical technique for mapping the receptive fields of the retina, regions that actually include more than one receptive cell, is to insert a microelectrode into a cell farther along "the visual pathway . . . to find out how we can most effectively influence its firing by stimulating the retina with light" (Hubel and Wiesel, 153). The receptive fields of the retina are mapped "systematically"—so as to preserve a "projection" of "the world they represent," though with some "distortion" in favor of "the central part of the retina" (150–152)—into the retinal ganglion cells, then into the neurons of the lateral geniculate nucleus, and finally into the neurons of the primary visual cortex. These latter cells, the cortical cells, are also divided into levels. In a typical experiment, a microelectrode was inserted into a "complex" cortical cell at the second level, after the level directly connected to the lateral geniculate nucleus. The retina was then stimulated with light in a variety of ways, and the first event that the experimenters recorded occurred at the designated cell:

> A typical cell responds only when light falls in a particular part of the visual world. . . . The best response is obtained when a line that has just the right tilt is flashed in the region or, in some cells, is swept across the region. The most effective orientation varies from cell to cell and is usually defined sharply enough so that a change of 10 or 20 degrees clockwise or counterclockwise reduces the response markedly or abolishes it. (It is hard to convey the precision of this discrimination. If 10 to 20 degrees sounds like a wide range, one should remember that the angle between 12 o'clock and one o'clock is 30 degrees). (154)

So, how shall we describe the connection between the firing of a typical cell and the event at its target?

We know, first of all, that David Hubel and Torsten Wiesel did so in classical terms. They spoke of the "best response" of a typical cell as if it were an effect of a stimulus at the retina, making it the last event of a causal chain from the source of the stimulus to the designated cell. This causal chain is then said to instantiate a certain computational or logical "scheme," a "transformation" of "incoming information" so that the designated cell "responds not to spots of light [as do the preceding cells] but to specifically oriented line segments" (154). As for how the "transformation" itself is accomplished, Hubel and Wiesel concluded that the "simplest" of the cortical cells in question, for example, "behave as though they received their input directly from several [preceding] cells with center-surround, circularly symmetric fields," whose centers are "all excitatory or all inhibitory, and lie along a straight line" (154). While Hubel and Wiesel lamented the tradition of taking the representative character of the cortical maps to be "an end in itself" (152), their imagination was nevertheless often captivated by the character, which is especially useful on an outside-in approach to the visual nervous system.

Their theory of visual nervous system activity, then, is certainly more sophisticated than Newton's, but only in detail, not in form. (Just replace Newton's vibrations with the appropriate physical events.) Each event in the series of events in question is still taken to be separable from the events that follow it. Or again, each event is already definite as the cause of the event that follows it: each cell is waiting to be stimulated by an event that is already constituted, namely, the firing of certain preceding cells. Hence, the specification of each event does not depend on the specification of the events that follow it. Even given the experiment of Hubel and Wiesel, the specification of the very last event at the designated cell does not depend on the specification of the events that follow it: "There is no direct evidence that orientation sensitive cells have anything to do with visual perception" (155). Hubel and Wiesel must have slipped, therefore, when they referred in the above quote to the work of the designated cell as "discrimination" (154). After all, they intervened in visual nervous system activity to put *themselves* in a position to infer a connection between the firing of the designated cell and the event in its receptive field; they specified both events *before* they inferred the connection between them. How may the visual nervous system itself be said to infer the connection in question? On the approach of Hubel and Wiesel, it may only "respond" at the next level of its arrangement to the firing of the designated cell, and the form of the next level of its arrangement cannot differ from the form of the preceding levels: the firing of *any* cell is the effect of the firing of the relevant preceding cells, constituting what is known as a neural circuit. (Is it any wonder, then, that Descartes, Newton, and Bohm all resorted to something outside of the series of events in question to resolve an inner show?)

On the other hand, thinking of the experiments of Hubel and Wiesel as versions of the additional thought experiment (so that the source of the stimulus is the target, for example), we may express the physical situation in terms of wave functions through the first chemical events in the relevant region of the retina, and continue to do so until we reach the event at the designated cell, which we "directly observe." We are in a position to say what the definite properties of the preceding events are only on the basis of our observation of the event at the designated cell, but moreover—and here is the difference that being in the quantum context makes—only with the event at the designated cell does our description leave the level of mere potentiality. (On analogy to the classical version of the original microscope experiment, if Hubel and Wiesel had been in a position to say what the definite properties of the preceding events were only on the basis of their observation of the event at the designated cell, the resulting description would still have left the level of potentiality at the very source of the stimulus.) In the quantum context, the firing, indeed, the "actualization," of the designated cells is "entangled" with all the preceding events, including the one at the target in "the visual world"; the link particles must include not only the relevant photons, but also the relevant particles between the receptive cells and the designated cell, thereby extending "the whole" deeper into the visual nervous system. The designated cell is, let us say, the site of the first chemical event of the track in *this* experiment once we describe the experiment at the quantum mechanical level of organization of body. (Compare this experiment to one in which we insert an electrode in any preceding cell, including a receptive cell.) Like other experimenters who did not place their experiments in the quantum context, Hubel and Wiesel proceeded as if no order of description of body were to entail an *internal physical connection* between the events in question: all of these events are aspects of "one whole."

In the first version of the additional thought experiment, we *pretend* that we *need* to let the description of the first chemical event serve as the means for drawing a line to separate the observing instrument from the observed object, and this kind of pretense is found throughout the practice of quantum mechanics. Even a question as to how we know where to draw a line is misleading if we take it to imply that a line has somehow already been drawn without us? Or again, as Bohr put it (137–138), where we draw a line is "a free choice." (Here we must be clear that we are reaching these conclusions while approaching the first observer's visual nervous system from outside in. What that system itself may be said, from inside out, *to do* in virtue of the chemical events in question is another matter altogether, which we will soon explore.) In our first version of the additional thought experiment, we *choose* to draw a line at the first chemical event, that is, we *take* it to be *already* definite in one way or another, exactly as we do when we refer to the point of entry into the emulsion in the original thought experiment—what Bohm called "P"—as if we were to "directly observe" it. The need to ask what the point of entry in the additional thought experiment *is*—or better, the need to ask what puts us in a position to *say* what the point of entry *is*—is simply much more obvious, though no less necessary, than in the original thought experiment. In a quantum context, then, Hubel and Wiesel *chose* to draw a line at the event at the designated cell, constituting the designated cell as the point of entry into the emulsion of their experiment.

Now, before we consider in what sense the visual nervous system itself may be said to draw a line, let us imagine a case in which we finally follow the approach of Hubel and Wiesel to its natural end, thereby tracing the series of events in vision to those cortical cells whose firing is said to be vision itself. (Here our imagination runs to more than one cell, but the case would not be fundamentally altered were it to run to only one cell.) Let us imagine as well that vision is at or near its threshold, making it scotopic vision. Here we will be confined to a case of "visual perception" as close as we can come to one in which the subjects simply "discriminate" a stimulus of the kind that Hubel and Wiesel considered: given a stimulus below the threshold of other vision (really, below that of foveal vision, which involves the cones, rather than the rods, of the retina), but above the threshold of scotopic vision, we ask the subjects, "Did you, or did you not, see that flash?"

At the same time, so we imagine, electrodes inserted in the appropriate cortical cells of these subjects reveal that those cells fire when the subjects say "yes" and fail to fire when they say "no." We have *chosen* to draw a line at the very last cells to be involved in this case of vision; the emulsion in *this* experiment begins at the site of these cells. Hence, the events at the designated cells are "entangled" with all the events that precede them, stretching all the way back to the source of the stimulus. All the events, again, are aspects of "one whole" that cannot be analyzed into autonomous components. Is it any wonder, then, that in an experiment of this kind, as E. Baumgardt concluded (35), "the uncertainty of seeing is predominantly due to fluctuations in the actual numbers of quanta absorbed by the retinal rods?" If the relevant nervous system activity after the first chemical events, or even after the events at the cells studied by Hubel and Wiesel, were essentially separable from the first chemical events, how do we explain that the uncertainty of seeing is so closely connected to the first chemical events? (Hubel and Wiesel themselves stressed that, "whatever any region of the cortex does, it does locally" (152).) It is interesting to speculate that the wholeness of scotopic vision arises because vision in general, even the nervous system as a whole, *works* at the quantum mechanical level of organization of body.

After all, if we can use the visual nervous system as an instrument at the quantum mechanical level of organization of body, why not suppose that the visual nervous system can use itself as such an instrument? If we study the visual nervous system from outside in, where we draw a line to separate the observing instrument from the observed object is a free choice, but if instead we study the visual nervous system from inside out, we may look for evidence that the system itself draws a line. As a passive emulsion, exactly like the emulsion in the original thought experiment, the system is only capable of a response to a stimulus, albeit a response that turns out, in the quantum context at least, to be an aspect of "one whole" along with the stimulus. Still under the classical spell, we have not yet changed the basic form of our approach, as if the visual nervous system could not be said to use itself. If we are body inside out, however, such an assumption makes no sense. Fortunately, it also made no sense to other experimenters who were working outside of the quantum context, but who nevertheless thought of the visual nervous system inside out, and so looked for evidence that its "transformations" of "incoming information" were not entirely passive.

In fact, "as in the case of other sensory modalities, stimulation of the appropriate part of the cerebral cortex resulted in changes in the receptivity (for example, the size of the receptive field) of the retinal ganglion cells" (Pribram, 87). These changes are regarded as the direct result of a "feedforward" process that involves *the very event that follows the first chemical event of vision*, that is, *the very first neural circuit*. Specifically, the "feedforward" process alters the inhibitory interactions of the layer of *retinal* cells between the receptive and the ganglion cells:

> The sites where such influence is most likely to be exerted are, of course, the neuronal inhibitory interactions which occur in the input channels, the interactions which, when grouped as logic elements, compose the screens through which and onto which input is transmitted. By increasing lateral inhibition, for instance, sensory contrast can be enhanced and the recovery system slowed (since lateral inhibition is assumed to reciprocally influence decrementing). Slowing of recovery in the primary visual system is in fact observed when the inferior temporal cortex is electrically stimulated. (322)

Moreover, inasmuch as the effects of stimulation were observed only when the subject—a monkey—was "bored" and not "attentive," Karl Pribram concluded:

> Taken together, these experiments show that the effects of electrical stimulation of the inferior temporal cortex and those produced when the monkey is "attending" are similar, and that the two processes show a considerable amount of convergence onto some final mechanism. It becomes reasonable, therefore, to suggest that the process of attention involves the influence exerted by the inferior temporal "association" cortex on the input mechanism. Through this influence attention is able to alter the time course of inhibitory interactions in screens and thus the characteristics of the Image initiated by any particular input. (324)

Later (327–31) Pribram compared this process to the shortening or lengthening of a zoom lens, as if more or less of the "Image" were to be "in focus," though on this comparison we cannot help but wonder what plays the role of the eyes that resolve the image we bring into focus with an actual zoom lens. (In Bohm's analysis our awareness plays that role.) Only the properties of the "Image"—not even those of the input, let alone those of the target—were taken to be dependent upon the "feedforward" process of the visual nervous system. Pribram was still working, in other words, in the classical context.

To understand the significance of Pribram's work in the quantum context, let us first return to the version of the additional thought experiment based on the actual experiments of Hubel and Wiesel. We still insert an electrode only in the designated cell at the second level of arrangement of the primary visual cortex, and on the basis of the evidence to be gained thereby we make inferences about the target. However, now we know that if we continue to express the physical situation in terms of wave functions as we move from the receptive cells to the designated cell, we will have to include the appropriate reference to the entire "feedforward" process responsible for the very structure of the first neural circuits. The firing of the designated cell is thereby "entangled" not only with the series of events we specified before, but also with the entire series of events in the "feedforward" process. It is *this* "whole" that is resolved by our experiment, and it includes events at cells formerly regarded as the sites of events that occur only after the event at the designated cell. (Doesn't this make the wholeness of scotopic vision even more impressive?)

Now, secondly, let us interpret Pribram's result from inside out entirely. Just above *we* are still using the visual nervous system while at the same time taking into account how it is using itself. When we use it, we intervene by gaining access to the designated cell, but we do not wish to change the way in which the designated cell works, at least not with respect to the preceding events; we wish only to make possible, so to speak, a certain track of the event at that cell, the point of entry into the emulsion of our experiment. So, does the visual nervous system use itself in an analogous way? When it uses itself, the "feedforward" process reaches to the events which follow the events at the receptive cells, but does not alter the way in which the receptive cells themselves work, at least not with respect to the preceding events; instead, the process makes possible, so to speak, certain tracks of the events at those cells. The visual nervous system evidently intervenes in its own work exactly as it must do if it is to be said to use the eyes as they are used in the very first version of the additional thought experiment, that is, as Heisenberg microscopes. Why not say, then, that the visual nervous system itself draws a line exactly where we should have expected it to do, namely, at the very first chemical events of vision, thereby constituting the receptive cells as the points of entry into the emulsion of the eyes? And if we can be said to resolve certain "wholes" by using the eyes as Heisenberg microscopes, why not say as well that *the visual nervous system resolves certain "wholes" by using the eyes as Heisenberg microscopes?*

So, if we reconsider the very first version of the additional thought experiment, but allow the eye to be a part of a living visual nervous system, we may describe the experiment in two ways. Either we use the eye, in which case we monitor the firing of the receptive cell by inserting an electrode in it, creating a certain feedforward-feedback system: let us say, we *perform* the track of the firing of the receptive cell. Or, on the other hand, the visual nervous system uses the eye, in which case it monitors the firing of the receptive cell by its own feedforward-feedback system: let us again say, it *performs* the track of the firing of the receptive cell. In both descriptions, the point of entry is the receptive cell and a certain "whole" is said to be resolved, but in the latter description "the whole" includes the entire visual nervous system, inasmuch as the track of the event at the receptive cell cannot be specified without specifying the entire feedforward-feedback process. That is to say, in both descriptions the relevant track can be specified only by referring to the experimental arrangement, and in the latter description that arrangement includes the entire feedforward-feedback process of the visual nervous system. From inside out, the system is an instrument that performs its own experiments, much as, evidently, Pribram imagined as well, though he did not interpret its performance in the quantum context.

Accordingly, the urge to continue to talk in terms of "an inner show" or "Image" separable from the events leading up to it must be resisted. Whatever show takes place, it cannot be thought of as "inner" as opposed to "outer": "inner" and "outer" are precisely the classical terms that our discussion undermines, by reducing their supposed referents to aspects of "one whole." Or again, in Newton's terms, the nervous system activity relevant to vision is not ultimately present to the soul; the soul's work becomes an aspect of "one whole" at the quantum mechanical level of organization of the nervous system and the physical environment. We speculate, in defense of Plato, that this whole *is* our visual experience of the world, in which we find *both* us and the world, *the very same human beings and world to which we refer in the-world-of-our-faces*.

Again, Plato believed his eyes, and therefore thought that the white thing he saw resolved in the world, not in some "inner show." Our work on quantum mechanics puts us in a position to believe our eyes once again. As against Berkeley, the visual nervous system regarded as an instrument can be said to resolve—Bohm used the term 'actualize'—the definite properties of the things placed in the target of the eyes. We speculate that as body inside out vision is not a trick played on us by our brains, as Crick believed, but rather, if a trick at all, a trick played on us by *both* our brains and the world. Indeed, how can it be a trick in the sense that worried Crick if the things in the world behave much as Plato said they do? Here Merleau-Ponty's epithet (1964a: 171) is most apt: "the metamorphosis of things themselves into the sight of them." Things themselves, let us say, wear the resolved properties, including the properties mistakenly taken by Descartes to be "inward." It is simply not always the case that the resolution or metamorphosis is successful, and when it is not successful we should no longer blame ourselves as if the world were not essentially to blame as well.

Now, inasmuch as our speculation is based on evidence derived mostly from studies of how monkeys and other beings who live in the-world-of-their-faces perform acts of "attending" (Pribram; Hubel and Wiesel), we must consider different ways in which visual nervous systems can work at the quantum mechanical level of organization of body. We must be able to take into account, at the least, that "attending" cannot work in the space of the world from a viewpoint as it does in the space-time of the world from a standpoint. When it comes to human acts of "attending," we must consider that a thing in the world can undergo a very special metamorphosis from merely being in the-world-of-our-faces to being of our-faces-in-the-world. Plato also spoke of a correlated metamorphosis of our own bodies: the eye that sees in the light of our-faces-in-the-world is not an "inward" eye, not the eye of mind or soul, but literally the eye of body itself. Contemporary accounts of vision no longer include just the eye of body; the metamorphosis they allow occurs after the eye of body has already done its work. Hence, these accounts are always under the strain of their own visual metaphors. Pribram, for example, spoke of the holographic reconstruction of images by the visual nervous system, though actual holographic reconstructions *resolve* only in the presence of a seeing eye. And what *can* play the role of that eye in the resolution of vision if the eye of body has already been left behind? (Again, as Bohm admitted, any sort of construction, including those of computational schemes, must be present to an awareness that is not a part of the construction itself; hence, this awareness plays the role of our eyes, and visual "attending" may in no way be interpreted as stretching beyond the construction, let alone all the way to the target in the visual world.) Plato evidently decided to stop the regress to further "eyes" before it could get started. If we follow him, how shall we understand the metamorphosis in question, especially when it comes to understanding the resolution of the near side of "attending?" Let us now turn to the task of answering this question, thereby setting the stage for our final analysis of the origin of inquiry.

VII. End of Inquiry

Attention is the *only* phenomenon we speak of that joins what is otherwise supposed to be divided without moving through that division; we cannot say that attention travels to its object. Attention is always already caught by its object, though—and this will become important later—attention can be turned to one object after another, and each turn must therefore constitute a kind of discontinuity or jump. Plato could interpret the stretching of visual attention to its object only in terms of the order of movement of light. But we can try to interpret it "below" this order of movement. Hence we speculate, as in the last section, that the quantum mechanical level of organization of body has this significance: a relation of attention *is* the quantum mechanical correlation of "one whole" resolved by the work of our visual nervous system.

To understand the force of this Platonic theory of vision, let us first return to the quantum mechanical context itself. Some twenty years after the original version of an EPR experiment, in which two particles were correlated in terms of their position and momentum (Einstein, Podolsky, and Rosen), Bohm thought up a version in terms of the spin states of two particles originally in a singlet state. He considered "a system composed of two one-half spin particles to be in a singlet state (total spin is zero) and its two particles to move freely in opposite directions":

> Once the particles have separated without change of their total spin and ceased to interact, any desired spin component of particle 1 may be measured. . . . The total spin being zero, one knows immediately, without in any way interfering with particle 2, that its spin component in the same direction is opposite to that of particle 1 . . . , however, quantum mechanics . . . allows only one of these components to be specifiable at a time. . . . (Jammer, 235)

We can say neither that the two particles are connected by a signal (we cannot violate the special theory), nor that they are correlated all along in the way that we discover them to be by measuring particle 1 at some later time (we cannot employ any of the family of local, realistic theories). Hence, physicists themselves often have recourse to a certain metaphor: the two particles act as if they were to *know* about each other (Bohm; Shimony). (Shimony (1982) coined the epithet 'to feel at a distance.' Simple versions of the double-slit experiment also illustrate correlations between particles that lead physicists to speak in terms of the same metaphor (Zukav).) But is this really a metaphor?

Our speculation about quantum mechanical correlations is based on our ability to come to know about the target particle, from outside in, by using an eye as a Heisenberg microscope. We suppose that the visual nervous system can also use an eye as a Heisenberg microscope. In both cases, the particles at the near and the far sides of "the whole" resolved by the use of the eye act as if they were to know about each other. But in the latter case such acting exactly matches what we say about the case from inside out: the near side of "the whole"—the relevant particles of the visual nervous system—act as if they were to know about the target particle. So, why not extend our speculation to the use of artifactual instruments as well? Physicists could have said that the two particles in a singlet state simply *heed* each other. The correlation between the two particles is, let us say, *a simple relation of attention*. Each of these relations is assigned a certain probability by the wave function that describes the two particle system, and only one of them may be actualized at any one moment. They are, formally speaking, ambiguities in the order of movement of light that constitutes the two particle system at the limit of inner movement. These ambiguities may be resolved by the work of artifactual instruments.

Bohm has spoken of the correlation between the two particles as a kind of fold of body. The separation between the two particles is understood in terms of space-time, but the correlation between them is understood in terms of "a higher-dimensional reality," through which the two particles may be folded together: each particle "acts as if it were a projection of a higher-dimensional reality" (Bohm, 1980: 186–9). (More recently Freedman and van Nieuwenhuizen have reported work on "eleven-dimensional structures" that "might give a unified account of the four basic forces of nature," as well as of "quantum-mechanical effects" (74–81).) Let us say that such a fold of body is *simple*, but at the same time let us note that the body in question extends "below" the limit of inner movement. What, then, is a simple fold of body from inside out? It *is* a simple relation of attention. Although the fold of "one whole" resolved by the work of our visual nervous system is certainly not simple, it, too, *is* a relation of attention. An awesome array of simple relations of attention must somehow be resolved together in "one whole" by the work of our visual nervous system. (Think here of each possible experiment of the type that Hubel and Wiesel performed with respect to the visual nervous system.) To grasp the nature of this resolution, we need to return to our discussion of Einstein and telling the time.

Notice first of all that looking at a clock to tell the time has exactly the same form as our additional thought experiment: in this situation the target is simply the hands of the clock. Hence, we cannot hope to describe the definite properties of the hands of the clock as if they were independent of our move to tell the time. Our move to tell the time and the properties of the hands of the clock must be thought of as aspects of "one whole," resolved through the work of our visual nervous system with the many photons that link the hands of the clock to the receptive cells of the retinas. Or again, an everyday context in which we move to tell the time is also a quantum context.

This context, moreover, is exactly that of Einstein's definition of the time of an event once we have synchronized the relevant clocks: "Under these conditions we understand by the 'time' of an event the reading (position of the hands) of that one of these clocks which is in the immediate vicinity (in space) of the event" (24). As we noted two sections ago, the event in question may even be that of the departure of the link photons from the hands of the clock at the "time" of the position of the hands of the clock that stands apart at once, in view. Now we must ask whether or not the work of the visual nervous system with the link photons can provide the basis for our everyday claims to view the position of the hands of the clock, that is, to *know* the "time." As we also noted two sections ago, however, Einstein was not celebrating this knowledge as much as he was our ability to synchronize an event with the hands of a clock. Because he took the context itself to be that of a point event in space-time, he could neglect questions as to the accuracy of our work with the link photons: he *defined* the context as that of telling the time. He could not provide the ground for our actually telling the time because of the existence, from outside in, of a time delay between the departure of the link photons and our view of the hands of the clock. From outside in, indeed, we cannot avoid the skeptical question, "Does our work with the link photons actually constitute knowledge of the position of the hands of the clock?"

But compare this context to that of the additional thought experiment, in which, by our using an eye to measure the position of the link photon, we "know immediately" the position of the target particle; their correlation "actualizes" at once. The quantum mechanical way in which we claim to know the position of the target particle has exactly the same form as the everyday way in which we claim to view the position of the hands of a clock: the position stands apart at once. Why not simply say that our visual nervous system is an instrument that measures the positions of the link photons, thereby "actualizing" a correlation between them and the position of the hands of the clock, indeed, between itself and the position of the hands of the clock as well? From inside out, then, this correlation *is* a relation of attention: the position of the hands *itself* stands apart at once, in view. The answer to the skeptical question is that the position of the hands of a clock that stands apart at once, in the view of an observer in the immediate vicinity of that clock, *is* the time at the clock.

The fold of body in this context is the key. The fold must be *across space-time* between the visual nervous system and the hands of the clock in such a way that the time delay we made so much of when we approached the problem of telling the time from outside in no longer has primary significance; it must be said to come along *with* "the whole" resolved by the work of our visual nervous system. The same must be said about the space-time separation involved in the experiment with the two-particle system: it too must be said to come along *with* "the whole" resolved by the work of an artifactual instrument. Were that instrument actually able, from inside out, to make *use* of the simple relation of attention it resolves, it could pay attention to particle 2 *itself*, *at once*. Our speculation is that the human body *is* just such an instrument, though no doubt it does not work merely at the level of simple relations of attention. By identifying the quantum mechanical character of telling the time, therefore, we have discovered that the ground for claiming to tell the time lies in the very relation of attention that is routinely questioned from outside in.

Again, the problem with saying that the visual nervous system constitutes a mere computational "scheme" is that the resolution of the scheme is in the "eye" of the experimenter: it is the experimenter, after all, who intercedes in the work of the visual nervous system in such a way as to *know* the correlation between a cell's work and the event in its receptive field. How does the visual nervous system know? Why not say it knows simply because the cell's work is an aspect of "one whole" that, so to speak, folds together the cell's work and the event in its receptive field? Our speculation cashes in the metaphor of discrimination here as it does in the quantum context: the work of the visual nervous system, from inside out, *is* the resolution of the visual world. Approaching a cell from outside in, Hubel and Wiesel could not help but resort to metaphors of experience, for the cell's connection to the visual world cannot be intrinsic to its work. Only as body inside out may its connection to the visual world be intrinsic to its work, constituting a certain correlation between the near and the far side of "one whole": a relation of attention.

As we move along the visual pathway we imagine, as did Hubel and Wiesel in their own way, that numerous relations of attention may be resolved together, as "one whole," through the work of one cell. A clock hand with a certain "tilt," for example, will occupy the receptive fields of numerous complex cells at the level of the visual nervous system we considered in the previous section, and each one of these cells can be thought of as the site of the entry into the emulsion of an experiment performed by Hubel and Wiesel. Each one of these cells is also tied into a circuit of preceding cells, each cell of which is in turn tied into another circuit of preceding cells, and ultimately into the initial circuit of receptive cells; and each one of these cells can also be thought of as the site of the entry into the emulsion of an experiment performed by Hubel and Wiesel. The wave functions at each turn express the potentialities at that turn, any one of which may be "actualized" by an experiment at the site of a cell at that turn. But if the experiment is at the next turn, these potentialities become aspects of the potentialities at the next turn. This sort of nesting process must continue through the entire series of events involved in our coming to know the position of the hands of the clock, a series that, as Hubel and Wiesel admitted, must extend well beyond the cells they studied. In the previous section we imagined a case of the entire series, and the potentiality "actualized" in that case will have nested in it all the relevant potentialities at each preceding turn of the visual nervous system. So, too, must we imagine the case in which the target is a clock hand with a certain "tilt" and we come to know its position, though in this case our visual nervous system is said to use itself. The potentiality actualized will have nested in it all the relevant potentialities at each turn of the visual nervous system; that is, numerous relations of attention will be resolved together, as "one whole," "a whole" that folds together the clock hand with the visual nervous system. Hence, the metaphor is again cashed in: the visual nervous system acts as if it were to know about the clock hand.

We must now realize, however, that any claim actually to tell the time has an immediate implication, inasmuch as, to paraphrase Heisenberg, it is "meaningless to speak of the place of the hands of a clock with a definite velocity" (Jammer, 58). At the moment that we actually tell the time we cannot attach any meaning to the movement of the hands, *as if the clock could not be said to be at work*. If we claim actually to tell the time at a clock, we must also claim that the hands of the clock do not move continuously, no matter how the clock is made. A clock whose hands must always be said to be on the move is a clock on which we cannot be said to tell the time. Once we claim to *know* the movement of the hands of a clock, we cannot attach any meaning to a claim that the hands have a definite place; it is meaningless to speak of the movement of the hands at a definite place. Hence, whatever movement of the hands is claimed, that movement cannot be said to stand apart at once.

Again we have a perfect fit between everyday and quantum mechanical language. What kind of opening in experience must there be to claim to know movement if movement cannot be said to stand apart at once? As we remarked in part one, our original way of referring to an opening for action, to a moment, holds the answer: the opening must be dynamic, a "stretch" of the space-time of the world from a standpoint, the visual world. The opening for the view in which the position of the hands of a clock stands apart *at once*, however, cannot be a "stretch" of space-time at all. We may well correlate both ends of "the whole" resolved by the work of our visual nervous system: either both ends are squeezed to the limit of a definite place, or neither of them are, both remaining "stretches" of space-time. Only on the first alternative may we be said to view the hands of the clock at a place.

To interpret this result explicitly in terms of the work of the visual nervous system, we must finally pause to consider the application of a Platonic theory of vision to the resolution of the eyes themselves, the near side of "the whole" resolved by the work of the visual nervous system. In whatever way an eye is used as a Heisenberg microscope to measure link photons, the retina of the eye is evidently the crucial part of the instrument (though other "extra-retinal" parts, such as eye-muscles and the associated elements of the visual nervous system, are just as obviously required to complete the measurement if this system is using the eye). Bohr had said, of course, that the link particle and the crucial part of the instrument share the same fate in their description: if measuring definite place, a certain "uncontrollable" aspect of their interaction always eliminates the possibility of describing *both* in terms of definite movement, or vice versa; in the event of their interaction, either both are described in terms of definite place, or else both are described in terms of definite movement (135). Although this conclusion cannot be applied exactly in this case (because the eyes cannot be said to measure the momentum of the link photons), the nervous system must still be said to resolve either the definite place or the definite movement of the eyes, but not both at once.

Let us consider what it means to resolve the movement of the eyes. If the movement of the eyes is resolved, their place is not; it is meaningless to speak of the movement of the eyes at a place. If the place of the eyes is not resolved, the so-called construction of the visual world must proceed by a certain "averaging over the different arrays" of incident photons (Matin, 334, 349), and a crucial ability is that of varying the movement of the eyes so as to differentiate between their order of movement and that of the incident photons. That we have such an ability is not in question: eye movements called "versions" can "correct for velocity errors" (Alpern, 325) so that "ocular velocity matches target velocity" (Matin, 356). Especially crucial, in other words, is the ability to bring the eyes to rest, if only for a moment, relative to the target. One initial condition of the Heisenberg thought experiment is that the target particle and the emulsion are both taken to be initially at rest, at least relative to each other. (Whether the eyes follow a moving target or fix on a stationary target, the same experimental condition is obviously satisfied, though we must be careful to remember that any reference to the order of movement of the target that is not relative to the eyes, or vice versa, must be relative to some other reference body.) Moreover, even if the eyes cannot be said to be at a place, they can still be said to be at rest, or to come to rest, or to begin to move. All these are aspects of an order of movement, and may be resolved accordingly.

Now, the ability to bring the eyes to rest is not in question either, though it tends to be described as an ability to "correct for position errors" (Alpern, 325). This description is misleading if the eye may be said to instantiate only a certain order of movement. (If we say, for example, that the correction is *always* to match velocities with a target, then the case in question is simply the special case in which the target itself is at rest.) Bohm (1965) did an excellent job in discussing this order of movement. He concluded, not surprisingly, that the visual nervous system follows the principles of the special theory, aiming to abstract what is invariant from the set of variations that composes its primary data. These invariances are, as we may remember, certain *space-time* relationships. Again, the "position" of the eyes in itself is not a possible outcome of the work of the visual nervous system if the eyes may be said to instantiate only a certain order of movement. At rest or not, the eyes will always be situated relative to their target in an order of movement that encompasses all of them.

In this order, the movement of the eyes must be referred not only to the world, but also to the rest of the body. The eyes may be on the move, but the explanation for this movement may not involve the work of the eye-muscles at all, except perhaps to hold the eyes at rest relative to the head in which they are located: the eyes may be at rest relative to the rest of the body (or perhaps just relative to the head) but may be on the move relative to the target if the rest of the body itself is on the move relative to the target. For example, we may follow a moving target by matching head velocity with target velocity, though obviously the aim is to match eye velocity at the same time. The nervous system can never resolve the movement of the eyes without resolving, at the same time, a host of other movements. (If, for example, a cat grows up without being able actively to move, it is "hardly able to see" (Ornstein, 62).) And the fundamental movement here, as Gibson discovered, is along the ground: the ground, not our own body, is the reference body for the order of movement of vision. (It took us a long time *not to define* the Earth to be stationary, and even longer to realize that no reference body at all exists for light itself.) That is, in terms of posture, we are on our feet, at standpoints, as long as the eyes may be described only in terms of an order of movement. Here the eyes themselves do not take on any special significance when it comes to posture, as they obviously do when we view the world.

In the latter case, then, it must be the place of the eyes that the visual nervous system resolves. Again, if the eyes are said to be at a place, it is meaningless to speak of their movement, and so meaningless to speak of the movement of the rest of the body either, at least to the extent that speaking of it entails speaking of the movement of the eyes as well. Here the eyes may take on a special significance when it comes to posture, constituting the reference body of vision, with respect to which even the rest of the body of those eyes takes its place. It is no wonder that, as Rubin and Walls claimed about the "direction-giving cells of area-17," "each of these cells 'sees' as if it were in the retina in the position of the receptor itself" (360). Like Mathew Alpern in a quote above, Rubin and Walls referred to "the position of the receptor" without noting that we can speak of this position only if the visual nervous system resolves it, in which case the system works, as Rubin and Walls put it, from a "view*point*." Otherwise, we could say only that, for example, the eyes aim to be at rest relative to a moving or stationary target, that is, the eyes aim to establish a certain space-time invariance between themselves and the target. The behavior that Rubin and Walls tried to describe is really open to description at two levels, one of which is that of an order of movement and one of which is not.

The former order is precisely that of seeing the world. It is no coincidence that everyday accounts of this experience—and here I am thinking primarily of the account of the Hopi we discussed in part one—match its form exactly. We may develop a sense of place, but only if it is not separate from time or movement in general, and from our relationships to the various other occupants of our world in particular. We have a sense of place, in other words, only in space-time, in which we can neither stay at a place nor return to a place because any place is really a place-time. The primary characteristic of this order is, as the Navajo put it, "dynamic" (Pinxton, van Dooren, and Harvey, 15).

As soon as our eyes are said to be at a place, however, that place is separate from time or movement, and so too is the relationship of that place to the place of the target. (Notice again that the fold of body across space-time, though it exposes the position of the hands of a clock to view, does not thereby date the view.) Here the relation of invariance, so to speak, is of space, not of space-time. It is on this level of description of the eyes that the eyes can take on a special significance in terms of posture, allowing us to *begin* with a sense of place that is not dependent on any order of movement. Our speculation herein is that the ability of the eyes to "correct for position errors" is aptly described only on this level of description of the eyes, as the "versions" in question—the saccades—may aim at places or changes in places only if we have the ability to make a place at a point of view. Without this ability the saccades are, as we said above, able to aim only at a certain space-time invariance. We note that the saccades are capable of "returning the eyes to the starting point" (Alpern, 321), and that in "repeated saccades between two given points . . . the eye reaches a given point in every saccade at virtually the same interval after saccadic initiation" (Matin, 349). Only if we can say the eye is at a place can we make sense of these descriptions, especially the description of returning to the same place, which place must be defined separate from time or movement. To eliminate temporal relativity here, in other words, the visual nervous system must work beyond the limit of inner movement to resolve "one whole" in which both the near and the far sides are squeezed to the limit of a definite place.

We should not be misled by our discussion of telling the time, therefore, because telling the time is not the paradigm of viewing, despite the prevalence in everyday talk of claims to tell the time. The requirements for a view of the position of the hands of a clock are severe, to put it mildly. The paradigm of viewing is instead our experience of the background for the position of the hands of the clock, inasmuch as the face of the clock is read even if the position of the hands is not. Think here, for example, of the numerals that mark the twelve hours. For anyone who can tell the time, the numerals are always resolved in such a way that the face of the clock stands apart at once, in view. If we position ourselves in front of a clock so that, eyes open, we look directly at it, then whenever we open our eyes the face of the clock is read. Try not to view the face of a clock; that is, try to see it so that the numerals are not already read, exactly as young children see a clock before they can tell the time. (Notice that our view of the position of the hands of the clock entails that these very numerals are already read.) Indeed, try not to view the world; that is, try to see it so that any numerals, letters and so on—my favorite example in a class is the ever present EXIT sign—are not already read, exactly as young children see the world before they can read. Precisely this unwitting tendency to read the world is the paradigm of viewing.

We no longer experience the resolution of the ambiguities in the numerals on the face of a clock, a switch that must occur precisely as the switch between viewing the duck and viewing the rabbit occurs, *at once*. As we also noted in part one, the figure of a numeral is like the duck-rabbit figure, an ambiguous figure that is not resolved in the order of movement of light; the *difference* between the faces of the figure, one to the left, the other to the right, is not an aspect of the order of movement of light, at least not at the limit of inner movement. Or again, as Bateson might have put it, this difference does not make a difference at the limit of inner movement. There the figure does not have opposing faces; they resolve together, so to speak, thereby losing their status as faces. But if we do resolve the faces of the figure, we must have the ability to work "below" the limit of inner movement: as body inside out, we resolve ambiguities in the order of movement of light.

Let us consider again the ambiguity of definite place/definite movement: if light be the figure, so to speak, each side of the ambiguity is a face of the figure, and the two faces oppose each other, constituting an ambiguity that is not in need of resolution at the limit of inner movement. Or again, the *difference* between these two faces does not make a difference at the limit of inner movement; they resolve *together*, thereby losing their status as faces. So, if in the visual world, as we noted in part one, we must face the past because the present cannot be broken from the past, why not say that this break is the resolution of an ambiguity in the order of movement of light, in favor of definite place? At the limit of inner movement light always already moves at a definite velocity in every direction from any standpoint—the irreducible of the order of movement of light. "Below" the limit of inner movement, we discover that this is only one face of light, indeed, the face that opposes the face of light at a definite place: the price we pay for the celebrated irreducible of light is that light cannot be said to be at a definite place at either end of "the whole" resolved by the work of our visual nervous system. Only inside light may we find it at a place, so to speak, may we break the present from the past, allowing the world to stand apart at once, in view, and thereby as well telling the time—*precisely the origin of inquiry*. (Again, a fold of body across space-time allows what would otherwise be past to stand apart at once, in view, thereby constituting the present that is broken apart from the past.) Einstein's very definition of simultaneity—recall the specific phrase "an observer *placed* at the midpoint" (22, my italics)—eventually forces us to refer to an ambiguity in the order of movement of light that is appropriately resolved only at the quantum mechanical level of organization of body. Indeed, only inside light may we *resolve* the ambiguity in favor of definite movement either.

The same may be said of all the quantum mechanical differences, not just place/movement, for example, those of spin and polarization, the latter of which were employed in the most recent experiments with Bell's theorem. All these differences make a difference only "below" the limit of inner movement, but so far we have discussed only the case in which, from outside in, we use an eye to measure link photons with respect to the difference place/movement. What keeps us from imagining cases in which we use an eye to measure link photons with respect to other quantum mechanical differences? And in turn, what keeps us from drawing an analogy to the way in which the visual nervous system itself uses an eye?

It is certainly no mere coincidence that, in both the everyday and the quantum contexts, we cannot describe the face of a clock independently of the work of our visual nervous system. But herein we will not actually speculate that the quantum mechanical differences other than place/movement are involved in the resolution of such everyday ambiguities as the faces of numerals. What we will do, however, is to derive the ground for such speculation by studying in more depth how the visual nervous system can be said to resolve everyday ambiguities in the order of movement of light, especially those feats of resolution that cannot be understood merely in terms of the resolution of the ambiguity place/movement. We already know that to read the face of a clock we must resolve this ambiguity in favor of definite place. So we must establish the possibility that the resolution of the faces of numerals may be referred to an inside-out activity of the visual nervous system in such a way that speculation about the connection between this resolution and that at the quantum mechanical level of description will be clear, if not evident as well. The case to consider is that of "visual directionalization."

On the classical, outside-in analysis of vision, should the position or the movement of the eyes play any role in the construction of the "inner show," that role must be confined to feedback from the eye muscles to the relevant portion of the cortex. Such feedback is called "eye-muscle proprioception." However, according to both Pribram (88–90) and Rubin and Walls (364–67), "eye-muscle proprioception plays *no* part in visual directionalization." Rubin and Walls argued that the eyes must be actively, not merely passively, moved:

> If one stares at a bright spot of light, which is straight ahead, long enough to develop an after-image that can be observed for a while in the dark, one will find that finger pressure on the eyeball from any direction, with any amount of force, fails to "move" the after-image out of a subjectively straight-ahead position so long as one is "willing" the eye to aim straight ahead. But if the eye is "muscularly" turned, the after-image swings through space wherever the visual axis goes. . . . In the after-image experiment outlined above, the passive movement of the eyeball produced by the finger would stretch some muscles and stimulate their proprioceptors, but "nothing happens." (364–65)

Besides noting this difference between active and passive movement of the eyes, Pribram referred to an experiment in which an experimenter paralyzed his own eye muscles: when he tried to move his eyes "the visual world jumped in the direction of the intended movement," "independent of any change in muscular contraction" (88). Hence, eye-muscle proprioception is not the key to resolution of direction, which depends instead on an inside-out phenomenon, the very attempt to move the eyes, even if the attempt is an unwitting failure, as in any case like the one in which the eye muscles are paralyzed.

Rubin and Walls also created a fascinating thought experiment:

> Suppose that a person's eyeballs could be dragged forth from the orbits without breaking any connections essential for visual sensation—that is, the muscles and connective tissues would have to be cut, but the optic nerve and the blood vessels must not be. Having succeeded with this operation, let us rotate the eyeballs so that they are back to back. The left eye now aims left

> ward, and let us say that on its fovea there is imaged a table lamp. The right eye aims rightward, and on its fovea a little table is imaged. The two retinas are *anatomically* double-printed in area 17, and the lamp and the table are therefore *physiologically* double-printed there. The rules under which the visual system operates dictate that the "patient" *must* see the lamp on the table, and he must see both objects *in one direction*. The puzzle is, What direction will this be? Surely not the direction in which either object actually lies, for these opposite directions have exactly equal claims [that is, if considered from outside in] and neither could possibly be favored over the other. But if not the direction of either, then *what* direction—and why *that* direction and not some other? (363)

Well, given the evidence about eye-muscle proprioception *versus* intentional or voluntary movements of the eyes, they concluded:

> The lamp and the table would be seen wherever the active, "willed" innervations to the eye muscles placed them. The individual could not separate lamp from table, but he could see both wherever he pleased, simply by willing an appropriate voluntary eye movement. The fact that his eye muscles had all been severed would not be of the slightest consequence. (367)

Hence, this time quite strikingly, we must again understand the result in terms of an inside-out phenomenon, the very attempt to move the eyes, even if the intended movements never in fact occur.

Now, to continue the thought experiment, we must ask, does one *see* the lamp on the table or does one *view* the lamp on the table? In the attempt to move the eyes, does one attempt to move them with respect to the way in which one's face is referred to the world? Or alternatively, does one attempt to move them with respect to the way in which one refers the world to one's own face? Notice that the thought experiment is suited only to the level of the latter reference, inasmuch as the world of the experiment consists of the table and the lamp, and therefore the very foreground/background one uses to *see* is missing (for example, *the nose and the ground*). As their earlier remarks about directionalization also indicated, namely, that distance is laid off from "a view*point*" (or alternatively, from "the visual ego"), Rubin and Walls considered only viewing; their thought experiment does not make it possible merely to see in a direction. They expected us to think of willing the eyes to the left or to the right, and even employed these very terms in posing the experiment.

Again, were the attempt to move the eyes to be understood in terms of the order of movement of the eyes, it would be impossible to imagine how the ambiguity in question is resolved. The aim would have to be not only to bring the eyes to rest relative to the lamp on the table, but also to do so in at least two different directions. How could these directions be differentiated in the world of the experiment? If one is aiming to bring the eyes to rest relative to the lamp on the table (which is the only potential reference body other than the eyes in the world of the experiment), the direction of the eyes is already given. On the other hand, were the attempt to move the eyes to be understood in terms of the order of place of the eyes, it would be possible to imagine how the ambiguity in question is resolved: one's aiming would have the eyes themselves as its reference body, and the resolution of the direction of the eyes to the left, or else to the right, could be a function of the very attempt to move the eyes (without immediate reference to the lamp on the table). Indeed, *this* ambiguity of direction cannot be resolved in terms of the order of movement of the eyes even if the eyes move freely in their normal way, for the reference body of the order of movement of the eyes cannot be the eyes themselves.

Consequently, the thought experiment provides us with the key to understanding the resolution of the everyday ambiguities in question above, those that remain even if the eyes are said to be at a place. If the visual nervous system makes use of the quantum mechanical level of organization of body in order to accomplish this resolution, why not suppose that it does so precisely through the inside-out phenomenon in visual directionalization? We already know that an analogous inside-out phenomenon is crucially involved in the control of visual input (Pribram), making it not just another aspect of "the whole" resolved by the visual nervous system, but instead the aspect directly related to the resolution of the outside-in ambiguity concerning the point of entry into the emulsion of the visual nervous system. Why not suppose that further outside-in ambiguities in the order of the link photons require a similar inside-out resolution? From inside-out we resolve the place of the eyes as well as their directionalization; indeed, the two resolutions must be but one, for both are entailed by the relation of the view to the viewpoint.

If one suddenly wills the eyes from the left to the right in the thought experiment, an equally sudden switch must occur between viewing the lamp and table to the left and viewing them to the right. Does the thought experiment now remind us of anything? Is it not analogous to the situation in which one views the duck-rabbit figure against a blank background? Why not suppose that the subtle movement of the eyes that always seems to be involved when one switches from viewing the duck to viewing the rabbit, or vice versa, can be traced to acts of will analogous to those that figure in the thought experiment, in which the eyes, of course, cannot move at all? Do the eyes need actually to move in the case of the duck-rabbit? If the lamp and the table were each replaced by a duck-rabbit figure in the thought experiment, could an act of will analogous to the act of will in the original version of the thought experiment produce a switch from the duck to the rabbit, or vice versa, even though the eyes cannot actually move? (We must not be misled by the thought experiment to suppose that no metamorphosis of the eyes occurs just in the case they do not actually move; they are still inseparable from the inside-out phenomenon in visual directionalization, and at the level of organization of body in question we cannot describe the eyes apart from "the whole" which includes them as well as the inside-out phenomenon in visual directionalization. See the discussion of "attention deflection" at the beginning of part three.) Once again, we aim to move the eyes to a place, no doubt the center of focus (or of symmetry) for the faces in question, and at the same time to resolve a left-right ambiguity analogous to the left-right ambiguity resolved in the original thought experiment. It is hard to resist the conclusion that only one inside-out phenomenon is required: in resolving the place of the eyes, we automatically resolve the orientation of the view to the viewpoint in such a way that the orientation of the view to its center—a place, remember—is also resolved. Even if we do need to move the eyes from left to right in order to switch from the duck to the rabbit or vice versa—because the centers of focus for the two faces are different—the orientation of the face to its center, and of that center to the viewpoint, will still be resolved at once.

Upon the failure of an outside-in analysis of vision, then, a certain inside-out phenomenon becomes the key to the resolution of crucial ambiguities in light, making possible the duck-rabbit faces, left or right, and more generally, *all such faces*, including those of the letters, words, sentences and so on that constitute our writing and reading. The resolution of the ambiguities in question, so to speak, puts faces on the things in the-world-of-our-faces. The informal ambiguity in the figure of a numeral—the ambiguity as we could describe it in part one—belongs to the same family as the ambiguity at work in the thought experiment about the two eyes removed from their sockets. Both ambiguities require an inside-out resolution: we are able to attend to the relevant object in different ways, ways which we have mistakenly taken to be aspects of an "inner show." The visual nervous system actually works with the object in question to resolve relations of attention. The different ways we attend to the object are not aspects of an "inner show," but precisely ways in which *we stretch all the way to the object itself*. The show is as much "outer" as "inner," as the two ends of "the whole" are folded together in a relation of attention, inside the order of movement thought to separate "inner" from "outer." (It is as if the duck face and the rabbit face of the duck-rabbit figure were to play roles exactly like those of the place face and the movement face of the clock hand: the figure *itself* would then be said definitely to be its duck face or its rabbit face, exactly as the clock hand *itself* is said definitely to be its place face or its movement face.) The inside-out phenomenon that resolves informal ambiguities is precisely the same phenomenon that resolves formal ambiguities: *as body inside out, what stands apart at once in view is the world itself.*

In both viewing and seeing a relation of attention is resolved—the ends of "the whole" are "actualized" *together*—though only in the former case does the resolution reach its natural limit: the space of the world in view. In both cases, of course, we still enjoy the transparency of space-time from "below" the Newtonian level of organization of body, at the limit of inner movement. This transparency must be thought of as an outside-in transparency once we work from "below" the limit of inner movement; it constitutes the perfect partnership with the inside-out relations of attention, allowing vision to achieve a resolution unrivaled by the other senses even when that resolution does not achieve its natural limit. In the visual world, the ubiquitous transparency of places is not resolved because attention connects the movements at either end of "the whole" in question as if they were aspects of *one movement* (rather than of *one space*), thereby reinforcing the spatial-temporal relativity between the eyes and the target that we discussed above—the likely source of the resonance between the Desana's brain and the sky, and also of the tendency to emphasize process over substance by all peoples who stand on their feet in the visual world. Only when we understand that we work "below" the limit of inner movement, however, do we discover the foundation of vision in general, indeed, of experience in general, not merely of views in particular. A nervous system works so well because from inside out it *is* experience of the world: *to experience the world is the very nature of body inside out*.

We started this part with an outside-in approach to body that we developed from its classical to its relativistic form. On this approach we follow light from some target, say, a clock, until it strikes our retinas, and then we follow the nervous system until, presumably, a visual experience of the clock occurs; the experience is the last of a series of events, each of which is in turn the cause of the next event. In this way we aim to construct the experience from outside in, though we always seem only to push the experience to the limit of the construction. Or again, it is as if, on an outside-in approach, we were not able to get inside the experience in order to know how to construct it, exactly as we cannot get inside light in order to know how to construct it. We reach the limit of inner movement only to discover that we cannot get inside visual experience any more than we can get inside light. Both appear to be irreducible primitives, somehow mysteriously connected to each other.

But now, given our work on quantum mechanics, we understand the mystery: to get inside *both* we need to penetrate "below" the limit of inner movement. Inside light and visual experience we find the *same* relations. Or again, relations of attention may be said *to be* the correlations of "the wholes" resolved by the work of our visual nervous system. Inasmuch as relations of attention are inside-out phenomena themselves, moreover, quantum mechanics may be said to break the spell of approaching body from outside in. Indeed, quantum mechanics makes sense, that is, does not violate the special theory, only if we approach body from inside out: from outside in separation is overcome only by movement, whereas from inside out separation is always already overcome. (If one insists on trying to interpret quantum mechanics on an outside-in approach, one must refer to superluminal orders of movement without thereby violating the special theory, and the typical restriction is that such an order of movement cannot constitute a signal.) Attention connects but does not thereby move at all; it cannot constitute a signal, nor can it carry information, at least not in the sense in which information is not always already carried and therefore in need of being carried through space-time. "Below" the limit of inner movement no such information may be said to exist. Let us say instead that, as far as space-time is concerned, a relation of attention simply *is* information. (Again, as Bateson might have put it, a relation of attention *is* a difference that makes a difference. We do not communicate *by* attention: it is *always already* communication, though we must be careful to remember that we may speak of communication here only from inside out, not from outside in.) A relation of attention is, therefore, a primitive, an irreducible.

Faced with the problem of our inability to measure movement relative to the ether, Einstein realized that the problem should not be thought to be resolvable in the way that other problems are resolvable. The ether always drops out of the picture afforded by our measurements because we can never measure a difference in the velocity of light in different directions. So, why not *begin* physics without it? Or equivalently, why not *begin* physics with the equivalence of all inertial frames, which equivalence could be posed in terms of an irreducible of the order of movement of light: the constant velocity of light in all directions in every inertial frame?

Now imagine a contemporary physicist faced with our inability to understand the "collapse of the state vector," or the "actualization" of "one whole," at the quantum mechanical level of description of body. A great debate rages as to whether or not "consciousness" must be involved in an "actualization," and to make matters worse this debate carries over into the age-old philosophical debate about whether or not we can make sense of referring to something that is supposed to exist independent of "consciousness." Such a "consciousness," as we noted again just above, always drops out of the picture afforded by our measurements, and we tend to respond by taking "consciousness"—like the ether—to be responsible for the results of our measurements in a way that allows it to remain elusive. So, why not take a clue from Einstein? Although Bohr did speak of a "free choice" in quantum mechanics analogous to the "free choice" in relativity theory, he did not extend the range of that choice far enough to realize a further analogy with Einstein's choice, namely, that we should *begin* physics with an "actualization," ***the origin of inquiry itself***.

The measurement problem in quantum mechanics is not really a problem at all; quantum mechanics has just run up against the origin of inquiry, and justifiably finds itself unable to penetrate beyond it. We imagined that we could resolve the measurement problem because we did not clearly understand the full range of application of quantum mechanics to the workings of our own nervous system. We imagined, still under the sway of the old Newtonian framework, that we could inquire further, as it were, getting inside the irreducible of quantum mechanics as we got inside the irreducible of relativity theory. But "wholes" at the quantum mechanical level of organization of body reach all the way through our nervous system in such a way that the very origin of inquiry may be understood to reside in them as well. Or, to put the matter in terms of "free choice," by *defining* the quantum correlation in *every* "whole" as a relation of attention we turn around the measurement problem in quantum mechanics exactly as Einstein turned around the measurement problem in relativity theory by *defining* the velocity of light to be constant in all directions in *every* inertial frame. Einstein eliminated the ether, and we eliminate "consciousness." Analogously, we make neither a supposition nor an hypothesis about the physical nature of quantum mechanical correlations, but rather a stipulation which we can make of our own free will to arrive at a definition of attention, indeed, of *measurement itself*. On the stipulation that *we are body inside out,* our inquiry *begins* with "actualization."

In part one we concluded, purely in the terms of posture, that once in our history we must have experienced in a together-step of face/world. Even without making the stipulation about "consciousness," we now know a way in which this conclusion can be expressed in the terms of the order of quantum mechanical description. Whether in seeing or in viewing, both ends of "the whole" are "actualized" *together*; we *always* experience in a together-step of face/world. But once we make the stipulation, the conclusions of parts one and two are identical, not at all some kind of coincidence. As we will discuss in more detail at the beginning of part three, our experience and sense of ourselves arise in virtue of patterns of "actualization": *we* can be said to begin here as well. Although we have concentrated on inquiry with the world in view, we have noted that a rather different experience and sense of ourselves—visual selves as opposed to egos, for example—arise in virtue of the pattern of "actualization" that constitutes inquiry without the world in view, an inquiry which will be crucial to part three. Like our ancestors, we can conduct inquiry on our feet, though thereby we cannot reach the end of inquiry and the limit of resolution of attention in the space of the world in view.

We have already concluded that we began inquiry with a certain quality of experience, that of the space of the world in view. Now we must conclude that we *end* inquiry there as well. Relations of attention constitute the end of inquiry, in both senses of that phrase: we can approach body from outside in until we get "below" the limit of inner movement, and then we can approach body only from inside out, as what we *are*, not as an object of further inquiry. Even to say that "below" the limit of inner movement we will still approach body from inside out is misleading. We have no sense of ourselves at all, let alone as inquirers, except in virtue of "the wholes" resolved by the work of our nervous system. (To get inside of relations of attention, we must get beyond the very sense of ourselves, the very together-step of face/world, that makes inquiry as we know it possible; if the inside of relations of attention is an order of experience as well, as many meditational traditions claim that it is, then it is an order of experience without self, ego, or world at all. Again, see the discussion of "attention deflection" at the beginning of part three.) But, most importantly, it is precisely body—not "consciousness"—that constitutes relations of attention. We hereby arrive at a Kantian position without the original twist of idealism: body from outside in is, as it were, constituted by body from inside out. *This* body one cannot imagine oneself to enter into as Leibniz once did with respect to Newtonian body; it inspires instead an analogy with Aristotelian matter.

Although Rorty has very nicely traced the evolution of our conception of body, noting that Descartes broke with Plato by removing sensual experience from the body, Rorty himself carried on the Cartesian tradition by celebrating a "conversation" that is still alienated from body. The Cartesian split between mind and body is converted into a split between orders of description within the "conversation": one causal, non-holistic level and one non-causal, holistic level; the first for "explanation" in scientific discourse, and the second for "justification" in everyday discourse—or again, as Wilfred Sellars put it, "a logical space of causes" and "a logical space of reasons" (Rorty, 141, 389). Hence, our stipulation that *we are body inside out* is excluded from the "conversation," and along with it goes the opportunity finally to overcome the Cartesian tradition through an order of description of body "below" the causal, non-holistic level of its organization.

Like Rorty I am not urging a return to the old way of reaching outside the "conversation" to an independent reality and thereby constituting "objective truth," but unlike Rorty I am urging a new sense of objectivity based on an altogether differently ordered "conversation." In the terms of posture, we can understand not only the origin of inquiry and a host of other phenomena, but also how our "conversation," at its very foundation, is grounded in an order of co-making *with* the reality once thought to be independent of it, an order in which the skeptical questions as to what body is and how we know body simply do not arise. Only in terms of our being body inside out do we succeed in understanding our experience of the world. Such experience is the very nature of body, exactly as Plato thought. Or again, *we can trust our bodies*, exactly as the Greeks did. *Quantum mechanics is hereby understood as the science that turns common sense into the truth.*

Part Three

The Co-Making of Inquiry

I. Person

At the beginning of part one, we emphasized the contrast between being loose in an order of co-making and being free in an order of making, in a world in pieces. We supposed as well that our looseness had not broken us apart from each other and the world but was still an aspect of an order of co-making. This order could then be thought of as constituting our *common sense:* we were still bound together by some sort of common sense.

Parts one and two pursued this supposition on two, quite different levels of description. Part one set the stage for part two by using the terms of posture to create a discourse—a "mythos" or "true speech"—that concerned our evolution to the times of Plato and Aristotle. No attempt was made as yet to argue that this discourse constituted "objective truth," indeed, any kind of objectivity. We wished to show how the terms of posture could unite, in one pattern of discourse, diverse source materials that might otherwise be taken to be separable, disciplinary discourses. Hereby we hoped to engage in a type of philosophy Rorty called "edification."

The edification had a point, however, which was to reveal a tension between two postures, a tension that reached its definitive form, at least for the purposes of part one of the present work, in the theories of vision of Plato and Aristotle. The posture of viewpoints had no sooner emerged in our evolution when it led us to question the primacy of movement in our understanding of the world away from our faces. But as the space of the world in view came to dominate our experience, so too did the lack of evidence for our participation in an order of co-making: we seemingly did not stretch *to* the world.

Part two then took up the challenge to extend the pattern of discourse of part one, that is, the terms of posture, to cover the history of inquiry into the nature of body from Newton to today. Here again the edification had a point, but a point that eventually allowed us to rediscover the evidence for our participation in an order of co-making: we *do* stretch *to* the world, much as Plato (and Rolling Thunder) thought we did. The space of the world from a viewpoint is just as much an order of co-making as is the space-time of the world from a standpoint. In part two, in other words, we went beyond mere edification to create a pattern of discourse about body that unites physics, even today's physics, with the "true speech" of common sense, thereby constituting a new sense of objectivity.

Now, on this basis, part three must take up the challenge to extend the pattern of discourse of parts one and two, that is, the terms of posture, to the social history that parallels the history of inquiry into the nature of body from Newton to today. Although we do not engage in everyday life in the belief that our experience is a trick played on us by our brains—and at this level our common sense is still evident in our trust of our bodies—everyday life is organized as if we were free in an order of making, in a world in pieces. Not only do we fail to celebrate our looseness in an order of co-making. We fail to celebrate co-making at all turns, even when it is a question of realizing each other and the world through movement.

Before we try to understand how this came to pass, we need to develop the concept of a person we discussed in the introduction. This development must begin with the work of parts one and two, and then extend that work in such a way that the social history to be told in part three will have an adequate foundation in the new sense of objectivity. Especially crucial is the relationship between the concept of a person and the order of co-making of inquiry.

First, let us return to looking at the face of a clock. We noted above that each view of the face of the clock begins at once, though we do not catch the beginning as such: each view is always already resolved, its beginning in need of recollection. Now we must consider the retention of each view, or rather, each relation of attention between viewpoint and view. We must be careful because for everyday experience the passage between views—or, more simply, *attention deflection*—works so smoothly that we do not notice the discontinuities or jumps. We need to stare at the face of a clock, and be mindful, very mindful, of each relation of attention thereby resolved in order to notice that we do not have one view of the clock. A simple blink of the eyes, for example, punctuates two relations of attention.

Again, if we follow Plato in thinking of the eyes of the body as the eyes before which our vision resolves, the eyes constitute our viewpoint of the clock, resolved through the work of our visual nervous system.

> Surprisingly, our eyes are ordinarily in constant movement—even when we fix on a point, small tremorlike oscillations of the eyes can be recorded. These and other movements are in some people large enough to be conspic-

> uous to others, but—and here is the paradox—the person with exaggerated movements is unaware of them until they are pointed out to him while he looks in a mirror. (Pribram, 54–5)

Other people, such as meditators, claim to become aware of such movement without the aid of mirrors, from inside out rather than from outside in. Stare again at the face of the clock, and try to take note of each subtle shift of viewpoint, the place of the eyes. The harder we concentrate, the more obvious these shifts become, and each one ushers in yet another relation of attention, yet another view of the face of the clock. Attention deflection is a constant together-step of viewpoint/view.

The significance of this together-step will be more evident if we return to another task, that of viewing the position of the hands of the clock. Stare again at the hands of the clock, indeed, at the second hand of the clock, but now try to take note of the movement of the second hand; that is, try to take note of the together-step of viewpoint/view. (Remember, here we are trying to view movement, not to see it.) The harder we concentrate, again, the more obvious this together-step becomes, and the more obvious it is that *movement cannot stand apart at once.* Were we able to view movement, or to have a moving viewpoint, that movement would take place in the space of the world in view, involving a continuous series of definite places—hence Zeno's paradoxes, which simply cannot arise with respect to seeing movement, or to having a moving standpoint, that is, in the visual world, because there we are always working with a stretch of space-time and the very different together-step of face/world in seeing. Our supposed experience of viewing movement, or of having a moving viewpoint, is only a phenomenon of attention deflection, a certain pattern of together-steps of viewpoint/view. (It is helpful to think of each view as a frame in a film, arranged one after another for viewing, so that attention deflection can project them in a seamless pattern.) Each viewpoint/view is always already retained to the extent that it may be retained, and such retention is at once, separable from time.

Consequently, the way in which views are woven into an experience of the world is the key to the different ways in which views figure in our lives. They need not even be woven only with themselves, as they quite often are today. In part one, for example, we spoke of chimpanzees and early human beings who were not able to make a place at a viewpoint as we do, but who were still able to refer certain aspects of the world, though not the world itself, to their faces. Even before the resolution of the visual ego and the space of the world in view, our ancestors enjoyed an experience that was not confined exclusively to the space-time of the world from a standpoint. Early human beings then advanced to a together-step of face/world in each step of which the world itself was referred to a face. Indeed, all that differs from these early human beings to modern human beings is the pattern of the together-step of face/world.

The visual ego does not arise with just any pattern, as again any long-time meditator will tell us. Recall that, in order not to read the world, we must not allow attention to resolve any faces in the world. Or again, inasmuch as to resolve a face in the world we must be able to shift the focus of attention from one definite place to another, the center of focus for that face, we must not allow attention to focus on *a* place, opening attention, so to speak, to the periphery of the visual field. (See Carlos Castaneda, and the discussion of Miyamoto Musashi's work in section two. Also consider the *Shri Yantra,* an apparently symmetrical figure that actually has no center of symmetry: staring at such a figure turns into a constant shifting of the focus of attention from the whole to the center and back again, and again, until ultimately we resist the shift to the center, opening attention as well (Evans and Fudjack).) At the same time *we* cannot sustain any sense of being at *a* place, as attention opens at *both* ends, precisely what must happen if our consideration of the visual nervous system as an instrument in the quantum context is apt. The near end of attention becomes diffuse over our entire body, typically turning on the body's center, which is also the center of the zazen posture in which we cannot tell the left side from the right side of our own body. Once attention opens at both ends, we break the together-step of face/world in viewing; we return to our feet and the very different together-step of face/world in seeing.

As soon as we understand the nature of attention deflection, in other words, we undermine the sense of a continuously existing, isolated ego to which we are so typically attached. What constitutes the visual ego—that is, in the terms of posture, the retention of a face—turns out to be a phenomenon of attention deflection, exactly as the viewing of movement turns out to be a phenomenon of attention deflection. Whenever attention deflection ceases, so also do we, so to speak. Hence, in all those cases in which the crucial object or scene before the eyes moves with the eyes (or vice versa), viewing is no more possible than seeing: it is as if attention were caught in one relation, undermining any sense of ego. It is no wonder that two of those cases are hypnotism (for example, keeping the eyes on a dangling pocket watch as it moves like a pendulum) and meditation (for example, keeping the eyes on a certain figure or object). Sensory deprivation also defeats attention deflection in a similar way, with similar results.

Attention deflection, again, can be so seamless that it goes unnoticed at *either* end of "the whole" resolved by the work of our visual nervous system, as if either end were moving continuously through the space of the world in view, or alternatively, as if either end were continuously at rest in the space of the world in view—hence the two ends stand apart at once *as if* they were *not* two aspects of an order of co-making. Precisely this quality of experience constitutes the origin of inquiry, and especially crucial to this origin, it must be stressed, is that we are normally not aware of attention deflection at *either* end of "the whole" resolved by the work of our visual nervous system. (Otherwise, in the resolution of the duck-rabbit figure, for example, the jump between viewing the duck and viewing the rabbit would be tantamount to the jump between seeing the duck and seeing the rabbit.) Now that we are aware of attention deflection, however, we can finally develop the concept of a person on its basis.

Given the concept of a person developed in the introduction, to engage in the practice of breaking out of reading the world and each other is to aim to be a person. On our feet we realize ourselves as persons by default: a person *is* visible. We also based the concept of a person on the order of movement of body as opposed to the order of reading of body because the former was self-evidently an order of *joint* movement whereas the latter was not self-evidently an order of *joint* reading. But reading the world *is* a together-step of viewpoint/view, an order of co-making. It is no wonder that Goffman discovered, as we noted in the introduction, that "it is as if perception can only form and follow where there is social organization" (1972: 382). The sense of oneself that arises at this level of description—"what can be read about the individual by interpreting the place he takes in an organization of social activity, as confirmed by his expressive behavior" (356)—*must be social: reading is jointly with each other and the world.* The new sense of objectivity applies equally well to sociology as it does to physics.

The problem with the current readings of our society, again as we discussed in the introduction, is that they constitute social roles that obscure the aleatory ground on which the roles are reversible. Off our feet in the way prescribed by our society, in other words, we may never realize ourselves as persons; we may take ourselves to be merely "children" or "adults," "students" or "teachers," and so on. But we can also never be entirely visible again, that is, without giving up the special abilities that came along with getting off our feet. Hence, the special question of this part: what will make it possible for persons off their feet still to realize themselves as persons? To the extent that we are not visible—so the answer goes—our readings of each other will not obscure the level of aleatory body/person only if we realize each other through viewing *exactly* as we realize each other through seeing. Today we tend not to *realize each other* through viewing, even though the deeper base of viewing is an order of co-making. Again, off our feet we may still realize ourselves as persons to the extent that *co-making is evident to us.* Realized through each other and the world, we are *co-persons.*

We can sum up this section by asking what kind of term 'good' is. Let us consider two senses of the term, one which refers to what is in the world away from a face, and one which refers to what is away from a face in the world. The former is surely all that we need to understand what is good to the bee and the frog. What is away from their faces in the world is always ready in the world away from their faces, thereby constituting their automatic sense of what is good to eat: they are automatic eaters. But we certainly are not, having a mouth as loose as the rest of our faces. For us the good away from our faces in the world is not always already, if at all, the good in the world away from our faces, thereby undermining an automatic sense of what is good to eat.

We can note in this regard that the term 'approve' has the root to-try-or-test, which in turn has the root good. The bee does not try when it alights on the nectar guide: without a question of referring the nectar to its face, it immediately eats. A human being can raise a question, can try or test, where a bee always already has an answer. Such tacit knowledge saves the bee from having to discover what is good for it, and from quibbling with other bees over what is good for them. And what is good for a human being? We always say that no more profound question exists, but at the same time we do not note the special character of the second sense of the term 'good.' It is ostensive in exactly the same way that the second sense of 'yellow' is: one must *put* one's face *in* yellow in the first sense. (See the beginning of part two.) Hence, we say, "Try it. You'll like it!"

We try to find features of a human face that we all share, so that we can refer the good of human faces to that ground. But, of course, should we take this search to be on the level of what is simply in the world away from our faces, we must automatically agree on what is good, as the bees do by default. And if we take this search to be on the level—the uniquely human level—of what is simply away from our faces in the world, we run into the dilemma of our looseness again. Isn't the way out here *exactly* like the way out in the case of yellow? If the world is referred to it, each and every human face simply constitutes the point of view of a person: *the uniquely human good is an aspect of an order of co-making between co-persons.* Only by *realizing* this result will it be possible to cease to struggle—between ourselves and between ourselves and the rest of the world—just enough to begin to let our common sense flourish, as it must once have done, before our looseness broke it into pieces.

The special task of part three, in other words, is to show that our common sense only appears to be broken into pieces and that it may yet flourish in us. Amidst all the supposed brokenness, we engage in a great deal of cooperation, such as in managing to stock our stores with the same food year round. So, too, do we engage in a great deal of cooperation in our communities of inquiry. We *can* live and conduct inquiry in a different way, one that allows such cooperation to become the basic feature of social organization. But we must resolve, as noted in the introduction, the thorniest of our current problems, that of trusting each other enough so that we may realize our ability to work together at the level of reading as well.

Part three will be divided into four more sections. The following section will elaborate the discourse of realizing each other and the world through movement. In the process we will also begin the social history of the last three centuries, at least to make clear how our looseness in an order of co-making was turned against itself so as to create a world in pieces in which we were essentially isolated from each other. Then the third section will extend the social history to today, though with an eye toward a different tomorrow. We will look at various aspects of everyday life, and in each case we will wonder about the significance of movement and co-making even while we are showing how the significance was lost. Finally, in the fourth and fifth sections we will extend the pattern of discourse of the whole work to a future in which we can act on our common sense without giving up what we have gained by concentrating so exclusively on our individuality. In both everyday life and our communities of inquiry, even at the so-called leading edge of research, we can learn how to read each other and the world in a way that promotes our mutual well-being *with* the world: through the co-making of inquiry we can practice—as we put it in the introduction by citing Foucault—a "new politics of truth" that displaces the current "discursive regime."

II. Sensuality and the Panopticon

Before we humans came to read silently, we read by speaking, and the words—figures of sound—did not stand apart at once. Reading was still an order of movement that excluded any sense of distance separate from time. We can easily tell how far we have come from this stage of human development by trying not immediately to read every word—figures of light, such as 'now'—in which we put our faces. For us now the third word in this sentence immediately stands apart at once, facing right at a distance, as n-o-w.

Only in the space of the world in view can the figure 'now' stand apart at once as n-o-w or w-o-n, and in the latter case it faces left at a distance, as if we were reading Hebrew. N-o-w and w-o-n constitute an ambiguity in the figure 'now' exactly like the ambiguity in the duck-rabbit figure, an ambiguity that is not resolved simply by the light that illuminates the figure. Or again, were we only to refer our faces to the figure, we could never read it. To read the figure we must refer it to our faces so that it can stand apart at once, facing left or right at a distance.

In the space of the world in view places stand apart at once. Places are independent of movement between them; our destination is already waiting to be reached at the very moment we begin to move to it. The Hopi language, on the other hand, completely excludes all such spatial metaphors. Our destination is associated with its own time—that place, that time—which in turn is understood in terms of the movement required to realize our destination. Places are aspects of movement in the world.

The Hopi have a sense of distance, not stretching in a line away from the eyes, but rather stretching along the ground over which they must move. They are not swept off their feet into the space of the world in view; they are precisely on their feet, at standpoints. Thoroughly dynamic, the world stretches away from a standpoint in both space and time: an order of movement in space-time. If we are on our feet in the world, our faces will not be loose from the world; they will compass the world, together-stepping with it. We will be caught up in the purely visual system that Gibson celebrated, the eye-head-brain-body-world system. We will *see* the world, not *view* it.

Again, once we humans only saw the world and then came to view it as well. I therefore take the ancient Buddhist advice against having views literally. So did the seventeenth century, Japanese *Kensei,* or sword saint, Musashi. He advocated "the gaze in strategy," in which "it is important to see distant things as if they were close and to take a distanced view of close things," and as well "it is necessary in strategy to be able to look to both sides without moving the eyeballs" (54–5). Musashi then advised: "Use this gaze in everyday life and do not vary it no matter what happens." He wished us *to see all the time.* It certainly isn't any wonder that in fighting with a sword we should see our opponents, realizing them through movement, especially of our sword. But to see all the time, throughout everyday life, is another matter indeed.

Later in the text Musashi again warned us not to "fix the eyes on details":

> I will explain this in detail. Footballers do not fix their eyes on the ball, but by good play on the field they can perform well. When you become accustomed to something, you are not limited to the use of your eyes. People such as master musicians have the music score in front of their nose, or flourish swords in several ways when they have mastered the Way, but this does not mean that they fix their eyes on those things specifically, or that they make pointless movements of the sword. It means that they can *see naturally.* (89, my italics)

If seeing naturally pervades everyday life, then every aspect of the world will be realized *through movement alone.* I call this "sensuality" so that it can be said that *once we lived entirely in sensuality.* Only when our everyday life in sensuality began to break apart did we need, like Musashi, *to advocate* sensuality.

Let us imagine a cat who smells an open can of tuna on the counter top above. She turns her eyes to the counter top, but she certainly does not stand apart from the counter top at once. Seeing naturally, she is not limited to the use of her eyes; she stretches to the counter top with her whole body in that way so familiar to feline admirers. To her the counter top must be the realization of a certain whole movement that both unites her with the counter top—an example of the co-making of movement—and makes possible her uniquely feline embarrassment should she actually fail to reach the counter top. She has an *entirely sensual* relationship to the counter top.

If we are in the cat's place, won't we tend to refer the counter top to our faces, thereby standing apart from it at once? Then any movement we might make to reach it will be excluded from view. Unlike the cat, we tend to ask a question: how much effort of our leg muscles here now will result in a flight of our body to the counter top there now, through the intervening places that also stand apart at once, in the space of the world in view? We must then read our flight by imagining a point of view away from which we resolve an ambiguity in the figure of the counter top and our body: the stages of our flight at each of the intervening places stand apart at once, facing from now to then—what we called "the spatialization of time." (It is helpful here to think of each stage as represented by a frame in a film, arranged one after another for viewing.) *These* stages are certainly not resolved simply by the light that illuminates the counter top and our body.

We may recall the plight of the father in the story we told at the beginning of section seven of part one. After driving somewhere on a sunny day, a father and his very young daughter return at night. The daughter asks, "Why didn't we drive home on the road with the sun shining on it?" No doubt the daughter, confronted with a jump to a counter top, must still act more like a cat than like her father, who certainly imagines a point of view away from which he gets a good view of the whole drive, each stage of which stands apart at once, facing from start to finish. In no way does he still feel united, through a whole movement, with the sunny road. Like the cat, the daughter lives in what we called "temporal ubiquity," whereas her father lives in what we called "spatial ubiquity."

Yet, even for the father, seeing naturally may not be entirely absent from everyday life. In my life, for example, I can at times play pool by seeing naturally. Lining up my cue stick with respect to the balls and the intended pocket, I stretch to the pocket as an aspect of the whole movement that realizes the ball in the pocket. I do not stand apart from the balls and the pocket at once, in which case I would need to read them, turning them into objects of interrogation by asking how hard to hit the cue ball, with what spin, and so on. The answers, again, would not be resolved simply by the light that illuminates the pool table; I would feel awkward, breaking what athletes call "concentration." Instead I just feel the right movement through my whole body. Should the ball not fall in the pocket I will be surprised exactly in the way that the cat will be surprised should it fail to reach the counter top. This is also the kind of surprise felt by a person who first *sees* the duck in the duck-rabbit figure and then suddenly *sees* the rabbit.

Movement in space-time is *only visible* and, following Musashi, we will call it "natural movement." If we are swept off our feet into the space of the world in view, we cannot move naturally. We break apart Gibson's visual system—what we may call, as we will be in a better position to recognize below, "the subject/object, or the mind/body, split." This simple absence of sensuality pervades everyday life today.

Now, to begin to set the stage for understanding this absence, we must reconsider a letter we already considered in section five of part one. In 626 B.C., a subject of an Assyrian king reports that "my eyes are fixed upon the king, my lord." Today this report strains our imagination because we tend to take it to refer to the use of the eyes alone, as if the subject had intended to keep the king in view by constantly referring the king to his eyes. If we suppose, on the other hand, that the subject intends to *see* the king all the time, then the possible movements in question are, as Musashi put it, "not limited to the use of the eyes alone," and the subject need not "fix the eyes on details." Accordingly, the subject must refer *all* of his movements to the king: the range of his possible movements must be circumscribed in some way by reference to the king.

Inasmuch as this circumscription is an order of sensuality, whatever a thing is, it is in virtue of being an aspect of one whole movement; it is the realization of movement, but movement relative to other movement in one whole movement—what we called "together-stepping" in part one. When we say that all of the subject's possible movements are referred to the king, we do not mean that the king displaces the ground as the reference body for those movements, but rather that somehow the king's movements and those of the subject are folded together as a pattern of together-stepping. So, think of the together-stepping of a bee and the sun, taking the ground as the reference body of their movements. Then draw an analogy between how a bee compasses the sun and how a subject compasses the king: they both together-step, though to a different drummer, as we say. In the times in question, the subject might well become disoriented upon being cut off from the king, who constituted the center of the world (Frederick), often instantiated through architecture in cities that radiated from his palace.

If a subject does not obey the king, he will be out of step with the king, perhaps failing at together-stepping altogether. Typically the subject is only "out of formation" but still "on course" (Laing, 1970), still counted among the members of the society in question. (Here we can draw the analogy to the way a bee may wean itself from the sun by employing the edge of a great forest. In straying away from the great forest the bee must risk needing the sun again; it may be "off course," not only "out of formation.") The king must feel the subject's disobedience as a real blow, challenging his very position in the whole movement of the together-stepping of his people. Indeed, as Foucault put it on the basis of his own evidence, "by breaking the law, the offender has touched the very person of the prince" (1979:49).

In an order of sensuality at least, to be so "touched" is an immediate sensation, and the king's response must likewise "touch" the subject. Although we at times feel something akin to such sensuality—I have in mind a family in which, by breaking the laws of the parents, a child offender touches the very persons of the parents—our society does not constitute what Foucault called "ascending individuality":

> In certain societies, of which the feudal regime is only one example, it may be said that individualization is greatest where sovereignty is exercised and in the higher echelons of power. The more one possesses power or privilege, the more one is marked as an individual. . . . The "name" and the genealogy that situate one within a kinship group, the performance of deeds that demonstrate superior strength . . . , the ceremonies that mark the power relations in their very ordinary, the monuments or donations that bring survival after death, the ostentation and excess of expenditure, the multiple, intersecting links of allegiance and suzerainty, all these are procedures of an "ascending" individuality. (1979: 192–3)

It is in such a society that punishment aims "to make everyone aware, through the body of the criminal, of the unrestrained presence of the sovereign" (49), as in a public torture, an example of which Foucault relates in grueling detail: a drawing-and-quartering of a subject in 1757.

The king's response to a subject's offense, again, must "touch" the very person of the subject, thereby exhibiting *a real omnipresence* of which we can feel today only quite narrow examples, certainly not stretching across society as a whole. Along with natural seeing comes what we may call "natural authority," the essential aspect of which is that the people realize themselves together as one whole movement, with or without special reference to one among them. That one and only one person receives special reference at the top of ascending individuality—or that the persons who do receive it do so once and forever—is not necessary, but merely a rather late form of natural authority. In this light I am thinking of the work of Leacocke, who emphasized that decision-making was dispersed in classless societies.

From 626 B.C. into the nineteenth century, we witnessed the inflection between classless and class societies, even though we were throughout the period still fundamentally on our feet in an order of sensuality. Toward the end of the period the mass of the people at the bottom of ascending individuality—those who were most strongly connected as aspects of the whole movement of their society, and who eventually constituted the lower classes of class societies, including "vagrants, false beggars, the indigent poor, pickpockets, receivers and dealers in stolen goods"—together possessed too much natural authority, and thereby threatened the prevailing order of society. It was, according to Foucault (1979: 15, 63), "the breaking up of this solidarity that was becoming the aim of penal and police repression" during "the great transformation," from 1760 to 1840.

In 1787, Jeremy Bentham wrote a series of letters to a friend in defense of what he called the "Panopticon, or The Inspection-House, Containing the Idea of a New Principle of Construction, Applicable to Any Sort of Establishment, in which Persons of Any Description are to be Kept Under Inspection" (37). The principle of construction or, as he preferred to say later, of inspection is extremely simple:

> Ideal perfection, if that were the object, would require that each person should actually be in that predicament [of being under the eye of the persons who should inspect them], during every instant of time. This being impossible, the next thing to be wished for is, that, at every instant, seeing reason to believe as much, and not being able to satisfy himself to the contrary, he should *conceive* himself to be so. (40)

And as to the advantages of the principle of inspection:

> I flatter myself there can now be little doubt of the plan's possessing the fundamental advantages I have been attributing to it: I mean, the *apparent omnipresence* of the inspector (if divines will allow me the expression), combined with the extreme facility of his *real presence*. (45)

Bentham knew that he was stepping on toes, but only those of God and His "divines," not those of the king, who apparently could not boast of a sensual together-stepping with his people. On the side of the inspector, Bentham advocated "seeing without being seen," but also doing it often enough to create, on the side of the person under inspection, "the *feeling*" of omnipresence (44).

To grasp the "simple idea of architecture" (66), imagine a series of rooms arranged in a circle on each floor of a cylindrical building. Put a window in the outside wall of every room, but do not break the side walls, and let the inside wall be composed completely of "iron grating," including a door opening onto a circular walkway that connects the rooms to each other. (Bentham designed partitions that jutted into the walkway for some distance between rooms so that the prisoners could not see each other.) Finally, in the doughnut-like opening in the center of the cylindrical building, imagine an inspection lodge that commands "a perfect view" of the surrounding rooms, but that allows the persons in the rooms to see as poorly as possible into the lodge:

> Of all figures, however, this . . . is the only one that affords a perfect view, and the same view, of an indefinite number of apartments of the same dimensions: that affords a spot from which, without any change of situation, a man may survey, in the same perfection, the whole number, and without so much as a change of posture, the half of the whole at the same time . . . and that reduces to the greatest possible shortness the path taken by the inspector, in passing from each part of the field to the other . . . And I think, it needs not much argument to prove, that the business of inspection, like every other, will be performed to a greater degree of perfection, the less trouble the performance of it requires. Not only so, but the greater chance there is, of a given person's being at any given time actually under inspection, the more strong will be the persuasion—the more *intense,* if I may say so, the *feeling,* he has of being so. (44)

It is no accident that Bentham stressed that "the whole scene opens instantaneously to view" (46), or that, "confined in one of these cells, every motion of the limbs, and every muscle of the face [are] exposed to view" (47). He imagined that, so far as the field of inspection goes, the persons under inspection will stand apart at once in the view of an inspector. The panopticon could have been designed only in the space of the world in view.

Again, the "fundamental advantage" of the panopticon is "the *apparent omnipresence* of the inspector" (45). The panopticon is "an engine" of "such power" (141) that prisoners will assume their own inspection. If they assume their own inspection, they must feel *simultaneously isolated* from each other and thereby constituted prior to any movement to overcome their isolation. (If inspection were to isolate them in turn, as an order of movement of the inspector, their isolation could not be simultaneous.) "To the keeper," then, they are "a *multitude,* though not a *crowd;* to themselves, they are *solitary* and *sequestered* individuals," even "indulged with perfect liberty within the space allotted to them" (47). Whatever "*overt* acts" (66) arise through the liberty in a cell, they are always *in* the view of an inspector. (It is sufficient, as Bentham knew so well, that one's allotted space be simultaneously isolated from others *by inspection alone.* Opaque partitions between persons under inspection are not necessary if the liberty within one's allotted space does not include, say, setting eyes upon each other.) Each prisoner adopts what I call "a cell-like posture."

How do we elude inspection in the view of an inspector? Our acts cannot give away who we really take ourselves to be, especially if we intend to escape. We try to keep secrets from the inspector, and to be successful, the secrets cannot be resolved in the world away from our faces, simply by the light that illuminates our cells. That part is relatively easy: the secrets must be invisible. The hard part is that the secrets must also not be read by the inspector.

What is not resolved simply by the light that illuminates our cells will remain an ambiguity in the figure of our acts, an ambiguity that can be resolved only by reference to a human face. At the beginning of each part of the book we have discussed some examples of these ambiguities that are not at the level of the duck-rabbit figure. Perhaps the most revealing one for our purposes here is the problem of what is good to eat. A bee is an automatic eater; it automatically extends its proboscis when it lands on "a nectar guide," which is, for a bee at least, resolved simply by the light that illuminates a flower. We humans, on the other hand, are not automatic eaters. We interrogate potential food by referring it to our faces, and the answer—good-to-eat or bad-to-eat—cannot be resolved simply by the light that illuminates it. Nevertheless, by reading the figures of our acts of eating, an inspector may well discover how we refer potential food to our faces. Imagine all of our acts of eating to stand apart at once, in the view of an inspector, so as to reveal a pattern, as if that pattern were the duck or the rabbit in a duck-rabbit figure.

One way for us to keep a secret as prisoners, therefore, is to try to act as if we were acting automatically, in terms of what is resolved simply by the light that illuminates our cells. It is, of course, precisely our trying which is the basic secret, and it must not be evident in the figures of our acts. But Sigmund Freud said, remember, that no one can keep a secret. The ambiguities in the figures of a person's acts can always be read by an appropriately trained inspector. The plight of a person who assumes inspection is clear: the permanent possibility of being read by an inspector.

Now consider the positive side of this game. Other things equal, prisoners know that they will be released only when their sentences expire. But perhaps in the meantime their acts will be read by an inspector as deserving of earlier release. Again, imagine all of their acts to stand apart at once, in the view of an inspector, so as to reveal a pattern, in this case a pattern of referring aspects of the world to their faces in a way that represents, as it were, their good intentions. (I have heard teachers speak about the problem of sorting out those students who pay attention in class from those students who do not do so. This is an example, no doubt, in which the liberty in one's alloted space excludes setting one's eyes wherever one wishes, though setting them in the appropriate places, at least as far as may be resolved simply by the light that illuminates the classroom, cannot in itself constitute paying attention.) Short of actual escape, a prisoner's acts cannot directly accomplish earlier release. To be sentenced to a cell under inspection is to be induced to adopt a posture in which one's own acts cannot overcome the sentence. On either the positive or the negative side of the game, the price is the same: acts are never directly efficacious, but rather efficacious only through the view of an inspector.

Consequently, prisoners under inspection are driven away from movement in two, interconnected ways. First, prisoners cannot remain visible. To be visible they must move naturally, but in a panopticon distance is always already constituted, between places separate from time, in the space of the world in view of an inspector. To move naturally is to break out of a panopticon, not to move within it.

Second, prisoners cannot remain viewable either, at least if they wish to refer the world to their own faces. It is, after all, to an inspector's face that the world must be referred in order to read it in the appropriate way, the way that confers efficacy upon their movement. Prisoners must practice the art of shamelessness, as Goffman (1961) put it, for they must never seem to regard their movement as crucial in itself. It is their willingness to manage their movement *in* view of an inspector that is crucial. Prisoners need not actually take themselves to be deserving of earlier release, but they must act so as to be read—or even to be seen—as if they were so deserving. Any reference of the world to their own faces must be hidden in an ambiguity in the figure of their acts, preferably so well hidden that it will not be read.

Given Foucault's evidence as well, we can say that at the dawn of the nineteenth century the new object of penal justice is not the body. "From being an art of unbearable sensations punishment has become an economy of suspended rights," in which the body, formerly the primary aim of punishments, becomes an "instrument or intermediary," and the primary aim is a "bodiless reality," "the soul": "the soul is the prison of the body" (1979: 11, 17, 30). To take what we really are to be "souls"—and here I am not claiming that this practice appeared for the first time at the dawn of the nineteenth century—is to remove ourselves from our bodies in the same two ways that we are removed from our movement under inspection. As "souls" we are never simply present in our bodies, but rather revealed only through the resolution of an ambiguity in the figures of our bodies—what Laing (1971a) called "a schizoid state," in which our true selves are never directly in the world. *The price of a "soul" is a cell-like posture.*

Without the terms of posture, Foucault was certainly right to speak of "a slackening of the hold on the body" (1979: 10): "the great transformation" redirected "penal and police repression" through the body to "a bodiless reality." This manner of speaking is indeed one that he derived from the very texts of the transformation, texts which obviously parallel the classical analysis of body we considered in part two. But like us he himself remained concerned with "forces and bodies" (217), thereby helping us to redescribe the transformation in our own, simpler terms as a metamorphosis of the hold on the body, from one posture to another.

To be sensual what we take ourselves really to be must be visible, realizable through natural movement, and as Vico put it, all body. The price is to be simply present, though not as bodies in the space of the world in view. The advantage is that we are never constituted prior to movement; only by movement do we realize ourselves with each other. The solidarity of those who move naturally must therefore pose a constant threat to the organization of everyday life if that organization is in space so as to favor those who wish to be inspectors of that space. *Hence it is precisely sensuality that demands inspection.* Not only must we not feel connected as aspects of one whole movement, but such a whole movement must disappear from the field of our imaginations, thereby eliminating any sense of natural authority.

Bentham did not look very favorably, therefore, upon "*a dark dungeon under ground*," "a place of *secret* confinement" (54). To the keeper, the prisoners would now be a crowd even if they were to be inspected, and to themselves, though sequestered, certainly not solitary individuals. The principle of dungeons constitutes an order of movement in which the keeper exacts "unbearable sensations" upon the prisoners. The place of confinement would not differ in any essential from any other place "but that of the chance it stood of proving unwholesome"; the prisoners would be "scarce under any control," in a place "favorable to infection and escapes" (54). Even a dungeon that failed to be a place of secret confinement could not be an inspection-house:

> I hope no critic of more learning than candor will do an inspection-house so much injustice as to compare it to *Dionysius' ear* [a dungeon constructed by Dionysius the Tyrant of Syracuse in the shape of a bell with an apex that opened into the palace above]. The object of that contrivance was, to know what prisoners said without their suspecting any such thing. The object of the inspection principle is directly the reverse: it is to make them not only *suspect,* but be *assured,* that whatever they do is known, even though that should not be the case. Detection is the object of the first: *prevention,* that of the latter. In the former case the ruling person is a spy; in the latter he is a monitor. The object of the first is to pry into the secret recesses of the heart; the latter, confining its attention to *overt* acts, leaves thoughts and fancies to their proper *ordinary,* the court *above.* (66)

As long as silence alone provided secrecy the prisoners could not be under inspection, especially in later times, such as those in which Bentham wrote, when they might well be able to write and read in silence.

Indeed, the distinction that Bentham made at the end of the above quote does not even apply in the times when people could not write and read in silence, when thinking went on between people, in the open. We must imagine that even in Bentham's times, when people might well be able to write and read in silence, they did not live their entire lives in the requisite posture to do so. Presented with the written word, they could easily have strained to read it in silence if not aloud, exactly as in the times of Dionysius' ear they would have strained to read it aloud if at all. Natural seeing could still have dominated everyday life, allowing a large mass of people to feel connected to each other as aspects of one whole movement, thereby in turn allowing them to feel their natural authority or solidarity. In a dungeon, solidarity is the rule, and it was precisely such solidarity—"promiscuous association" (137)—that Bentham aimed to undermine.

To play it safe, however, he must have assumed that everyone had the ability that he had, namely, to take up his body from outside in, to be invisible, allowing his "thoughts and fancies" to be out of the reach of a contrivance like Dionysius' ear. So how could we affect his "thoughts and fancies?" Inspect their expression in the world: "*overt* acts!" Foucault put it as follows:

> He who is subjected to a field of visibility, and who knows it, assumes responsibility for the constraints of power; he makes them play spontaneously upon himself; he inscribes in himself the power relation in which he simultaneously plays both roles; he becomes the principle of his own subjection. (1979: 202–3)

He makes a place, that is, in the space of the world in view as if his body were in a cell under inspection. He thereby eliminates any possibility of sensuality and its associated solidarity.

Now, before we turn in section three to the extension of the cell-like posture throughout everyday life, I want to admit that I have not always put the above result in terms of sensuality. It surprised me that I could have failed to understand how important these terms were to my work. But, then again, I have always been immersed in an everyday life that celebrated the cell-like posture. So, one day when the above result was in the back of my mind, I felt a profound longing to touch someone whom I had not intended to touch. The longing suddenly pervaded my experience; every muscle of my body was poised to realize the other person through movement. Then just as suddenly the longing passed as it occurred to me that *this* experience was precisely an opening to natural movement, to what I now call "sensuality." If I had not been prepared through my work, I would have thought instead that this experience was simply some form of sexual attraction to another person, thereby conflating, as we tend to do, sexuality and sensuality, and in turn radically narrowing the scope of sensuality.

The key here—and what should be kept in mind throughout section three—is that once we are in space *we cannot be realized by touch*. (We are *invisible,* after all.) Just recall the sense of presence in our bodies when we do long to touch someone as I did that day: a mutual presence to each other as aspects of one whole movement. Bentham was well aware of the need to control such touching in everyday life, and what better way to do so, as I put it in my terms, than to make persons *unrealizable by touch in general?* It is no wonder that the moments when we are actually realized by touch mean so much to us. Only in those moments do we overcome the isolation of being always already constituted *at a distance from each other,* in which case touching is contact of the limits of bodies, open to sundry readings.

III. Panopticon of Everyday Life: The Cell-Like Posture

"Power produces," claimed Foucault (1979: 194), "it produces reality," and "the individual and the knowledge that may be gained of him belong to this production." This power is not natural but rather "political," as Marx (1967) put it, and just as it produces individuals it must draw its justification from them as well. It is a power of *discipline,* "an indefinite discipline: an interrogation without end" that produces "the disciplinary individual" (Foucault, 1979: 227) and the associated "descending individuality." The panopticon "automatizes and disindividualizes power":

> Power has its principle not so much in a person as in a certain concerted distribution of bodies, surfaces, lights, gazes; in an arrangement whose internal mechanisms produce the relation in which the individuals are caught up. The ceremonies, the rituals, the marks by which the sovereign's surplus power was manifested are useless. There is a machinery that assures dissymmetry, disequilibrium, difference. Consequently, it does not matter who exercises power. Any individual, taken at random, can operate the machine . . . [and the] more numerous those anonymous and temporary observers are, the greater the risk of the inmate being surprised and the greater his anxious awareness of being observed. (202)

Moreover, if we apply the inspection principle throughout everyday life, we create "omnipresent surveillance":

> It had to be like a faceless gaze that transformed the whole social body into a field of perception: thousands of eyes posted everywhere, mobile attentions ever on the alert, a long, hierarchized network . . . [in which] the [disciplinary] individual is carefully fabricated . . . , according to a whole technique of forces and bodies. (214, 217)

Let us now turn to Bentham's own plan for the broad application of the inspection principle.

A century after Newton, Bentham's analysis of the work of "manufactures" obviously depended on the classical analysis of bodily activity into its simple components:

> In many species of manufacture, the work is performed with more and more advantage, as everybody knows, the more it can be divided; and, in many instances, what sets bounds to that division, is rather the number of hands the master can afford to maintain, than any other circumstance. . . . Many are the instances you must have found in which the part taken by each workman is reduced to some one single operation of such perfect simplicity, that one might defy the awkwardest and most helpless idler that ever existed to avoid succeeding in doing it. Among the eighteen or twenty operations into which the process of pin-making has been divided, I question whether there is any one that is not reduced to such a state. (51, 56)

And, once employers need not hire the more skilled "workman," other benefits come their way:

> "By the infirmity of human nature," says our friend Ure [1835], "it happens that the more skillful the workman, the more self-willed and intractable he is apt to become, and of course the less fit a component of a mechanical system in which . . . he may do great damage to the whole." Hence the complaint that the workers lack discipline runs through the whole period of manufacture. (Marx, 1977: 490)

But Bentham knew what to do all along:

> And to crown the whole by the great advantage which is the peculiar fruit of this new principle, what other master or manufacturer is there, who to appearance constantly, and in reality as much as he thinks proper, has every look and motion of each workman under his eye. (56)

The principle of inspection is a perfect fit for the "mechanical" system of manufacture, as the divisions of manufacture stand apart at once in the view of the "master or manufacturer." The aim is not detection but *prevention,* to create *discipline.*

As Braudel made clear, prior to the invention of the modern arrangement of manufacture, well drawn by Bentham above, the place of work "usually consisted of a series of workyards, and individual craft units, rather than factories with a rationalization of tasks" (393). Hence, the inspection principle could not be applied: workers did not stand apart at once in allotted spaces separate from time, as at any moment their work might well require them *to move* in ways that the inspection principle could not tolerate. If we wish to achieve the same extreme facility of inspection in the inspection-factory as Bentham admired in the inspection-house, then any change in the use of a work space, even any change in the use of a tool in a work space, must be regarded as the mark of an operation composed of simpler operations that can be performed in work spaces that stand apart at once:

> Instead of each man being allowed to perform all the various operations in succession, these operations are changed into disconnected, isolated ones, carried on side by side; each is assigned to a different craftsman, and the whole of them together are performed simultaneously by the co-operators. . . . But as soon as the different operations of a labour process are disconnected from each other, and each partial operation acquires in the hands of the worker a suitable form peculiar to it, alterations become necessary in the tools which previously served more than one purpose. . . . Manufacture is characterized by the differentiation of the instruments of labour—a differentiation whereby tools of a given sort acquire fixed shapes, adapted to each particular application—and by the specialization of these instruments, which allows full play to each special tool only in the hands of a specific kind of worker. (Marx, 1977: 456, 460)

Whereas before we had one tool with many uses that were separated by the movement of the tool itself, now we have many tools with one use each, so that they can be separated in work spaces that stand apart at once. What were once different phases in the order of movement of one operation in one work space, are now different operations in different work spaces that stand apart at once. Again the principle of inspection is a perfect fit for the "mechanical" system of manufacture.

But an inspection-factory is not merely an inspection-house. It is not enough to induce the workers simply to adopt cell-like postures, for in that case they would have, as it were, perfect liberty in the spaces allotted to them. The workers must adopt cell-like postures as if the only possible movements in their cells were the very movements required by the system of manufacture in question. Nevertheless, the system of individual movements cannot constitute one whole movement because their connections are "external" to them:

> Only a few parts of the watch pass through several hands; and all these *membra disjecta* come together for the first time in the hand that binds them together into one mechanical whole. This external relation between the finished product and its various and diverse elements makes it a matter of chance in this case as in the case of all similar finished articles, whether the specialized workers are brought together in one workshop or not. (Marx, 1977: 462)

And again, what of the different phases of the work of each specialized worker? Exactly as we could combine two operations formerly performed in cells that stand apart at once, we could break apart two phases of an operation formerly performed in one cell. It is as if the temporal separation between any two phases of an operation were not essentially a function of the order of movement of the operation at all: the operation can always be broken apart so as to be performed in cells that stand apart at once.

It is this temporal separation, in a time separate from space, that "penetrates the body and with it all the meticulous controls of power," resulting in "a machinery of bodies" (Foucault, 1979: 152, 156). Again we have to imagine a point of view away from which each phase of an operation stands apart at once, facing from the beginning of the operation to its end. (It is helpful to think here of the way in which the different phases of a machine's operation can be represented in two-dimensional drawings, and then tacked one after another on a wall for viewing.) Then give this view actual spatial form by designating work cells for each operation, and organizing the work cells so that they can be inspected at once. The workers are simultaneously isolated in their work cells; to move out of one's cell is to cease to work. No sensuality unites them as aspects of the whole movement of their work, together-stepping with each other directly to bring about the desired end.

Certainly Marx never tired of describing the resulting alienation:

> The habit of doing only one thing converts him [the worker] into an organ which operates with the certainty of a force of nature, while his connection with the whole mechanism compels him to work with the regularity of a machine. . . . Not only is the specialized work distributed among different individuals, but the individual himself is divided up, and transformed into an automatic motor of a detail operation. (1979: 469, 481)

Again, this transformation was not lost on the proponents of discipline, such as Adam Ferguson, who wrote in 1767: "Manufactures prosper most where the mind is least consulted, and where the workshop may be considered like an engine, the parts of which are men" (Marx, 1979: 483). Eventually, then, the ideal of "the simple physics of movement" was abandoned in favor of a more "organic" conception (Foucault, 1979: 156):

> Discipline is no longer simply an art of distributing bodies . . . , but of composing forces in order to obtain an efficient machine. . . . The individual body becomes an element that may be placed, moved, articulated on others. Its bravery or its strength are no longer principal variables that define it; but the place it occupies, the interval it covers, the regularity, the good order according to which it operates its movements. . . . This is a functional reduction of the body. But it is also an insertion of this body-segment in a

> whole ensemble over which it is articulated. . . . The body is constituted as a part of a multi-segmentary machine. (164)

This counterpoint of reduction and whole-ensemble constituted, for Marx, the crucial aspect of the contradiction of manufacture (or rather, of capitalist manufacture): "it only accomplishes the social organization of the labour process by riveting each worker to a single fraction of the work" (1977: 464).

When Marx described the positive side of the social organization of the labor process, he stressed that "specialized workers are special organs of a single working organism that only acts as a whole, and therefore can operate only by direct cooperation of all" of them (466). But when he described the negative side he stressed that "the specialized worker produces no commodities"; "only the common product of all the specialized workers . . . becomes a commodity" (475). Hence, "the manufacturing worker develops his productive activity only as an appendage of that [a capitalist's] workshop"; "what is lost by the specialized workers is concentrated in the capital which confronts them" (482). The wholeness of *this* social organization of labor does not run very deep (that is, to our feet), for it is immediately referred outside the labor process, and hence it is not through their work itself that the workers form a whole-ensemble. (Here Sartre (1976) would speak of "the common alterity" of the workers, inasmuch as their organization "excludes the relation of reciprocity.") Again, their own movement possesses no efficacy in itself: they do not make the whole directly through their movement. The whole is constituted by the external relations imposed upon their work by the demands of the capital, or the capitalists, they confront.

But the same result follows upon the application of the inspection principle by itself, for only through sensuality can the workers develop a wholeness that is internal rather than external to their work. If the labor process were one whole movement, the price for a capitalist would be the solidarity or "promiscuous association" of the workers, in which case they would experience a form of natural authority. Indeed, the price would be the same for anyone who expected to exercise control of the kind that Bentham advocated. It is not necessarily as a capitalist—and I think Foucault must have seen this—that one introduces the negative side of the social organization of labor. (Witness the contemporary Soviet Union!) One need only aspire to *political authority as an inspector* (though, as we will make clear below, one cannot be a capitalist without being an inspector). Again, one need only elaborate the social organization so as to achieve precisely what Bentham aimed to achieve in the inspection-house, thereby creating the inspection-factory.

Once industrialization extends into the actual uses of machines, especially as in the modern assembly line, which is a machine itself, "we have a lifeless mechanism which is independent of the workers, who are incorporated into it as its living appendages," rather than "the living mechanism" of the workers in manufacture (Marx, 1977: 548). (Recall one popular image of the assembly line in which each place on the line stands apart at once from a glass-enclosed office elevated above the floor of the line.) Here again, it is possible, without the intervention of inspection, to incorporate the movement of the machines into an order of sensuality including the movements of the workers as well. The resulting loss of "discipline," however, would not be confined only to a "confusion" of workers. The lifelessness Marx abhorred requires inspection not only to isolate each worker simultaneously, but also to grade each worker in relation to each other in the whole ensemble of work.

This disciplinary knowledge, as Foucault thought of it, went "beneath the division of the production process" to "the individualizing fragmentation of labour power," and "the distribution of the disciplinary space often assured both" (1979: 145). Or, in our terms, it is *readable space* that assures both: good-work/bad-work becomes an ambiguity in the figure of work, only resolved away from an inspector's face. Again, work in itself is not directly efficacious, and if, as Marx claimed, our very nature requires it to be directly efficacious, readable space must be an arena of alienation. (See Linhart.) Whether or not we follow Foucault in thinking of the mechanism of workers as "organic," no division in readable space is *essentially* a function of movement; only *a machinery of bodies* is possible.

Now, such a machinery of bodies may well be imagined to be the ideal for social organization beyond the factory. Not only did Bentham imagine that "hospitals" and "mad-houses" would employ the inspection principle. He also applied it to "schools." (Foucault pointed out, by the way, that Bentham was not alone in advocating the inspection principle, and that as early as 1751 military establishments applied a version of the principle (1980: 147).) Exactly as Bentham imagined the work-cell, he imagined the study-cell, in which "the different measures and casts of talents, by this means rendered, perhaps for the first time, distinctly discernible" (63). Moreover, thinking of the school as a "total institution" (Goffman, 1961), Bentham also imagined that the students would occupy sleep-cells:

> The youth of either sex might by this means sleep, as well as study, under inspection, and alone—a circumstance of no mean importance to many a parent's eye. . . . The notion, indeed, of most parents is, I believe, that children cannot be too much under the master's eye. (63)

Hence, they opted for "the master's omnipresence" (63), and "the school became a machine for learning," as Foucault put it (1979: 165).

We will certainly need to look more closely at Bentham's sense of what parents in his day wanted for their children. As Aries pointed out, Bentham's sense did not really "spread to the parents" until "the nineteenth century":

> True, in the nineteenth century, when it triumphed, this tendency would also correspond to a secret desire to postpone the dreaded triumph of puberty. Any such idea of sexual morality was completely foreign to the reformers of the fifteenth and sixteenth centuries. But the desire to treat all pupils as very young children does not always spring from a sexual taboo, a puritanical impulse. On the contrary it precedes it: it seems in fact to be a manifestation of the modern idea of distinguishing the ages in a society in which they had become confused. (1962: 164)

And indeed of distinguishing the "talents" in a society in which they had become confused, or rather, undisciplined. To understand the more fundamental issue of distinguishing the ages—the institution of childhood—we must first consider the inspection-school and the nature of the authority invested in the master.

Bentham imagined that to carry out the intended scheme one needs to build an inspection lodge that does not allow the students to see the master while at the same time they become assured that they are being seen by the master. He also imagined partitions or screens between the students, though the screens may be "as slight as you please" (63). Today we simply arrange the students in rows that face the teacher. The students stand apart at once in the teacher's view, but none of them enjoys a similar advantage. If we recall the subtle, or often, not so subtle, means of restricting the use of the students' eyes while they are seated in such rows, we may understand just how slight the screens may be. By the time young persons get to college, if not before, it is nearly impossible to induce them to break out of their cell-like postures, to exercise the liberty of their own eyes; and certainly just sitting in a circle together is not enough.

The spaces in which the whole education of young persons is to take place are already waiting for them, in most cases even before they are born. Imagine a layout of these spaces in a community, in which each space is broken down into rooms, with desks arranged in rows, each labelled with the name of the young person who studies there. Imagine too that one desk is larger, facing the rows of smaller desks, and labelled with the name of the older person who teaches there: away from the teacher's point of view each student stands apart at once, alone. No act on the young person's part is ever directly efficacious: good/bad, ready-to-move-on/not-ready-to-move-on, is an ambiguity in the figure of a young person's work that can be resolved only away from a teacher's point of view.

As Aries (1962) and Foucault (1979) pointed out in their own terms, the stages of a young person's development were gradually organized in the space of the world in view so that we could read a young person's progress directly from one's place in these spaces. Imagine a point of view away from which the stages of development at each of the associated places stand apart at once, facing from the beginning of school to the time of graduation—what we call "age-grading." (For example, first grade here, second grade there, and so on.) These stages do not remain at the level of general development either, but instead penetrate to the minutiae of the young person's life in school. Think of the typical daily lesson plan of a teacher: it too is organized so that the stages of the plan may stand apart at once, facing from the beginning to the end of the day. As in an inspection-factory, at each juncture of the stages of the plan the master may well insert an inspection, as if these stages were cells between which the young person may move but only should the young person's work be read to deserve it. The inspection-school is a machinery of bodies as well.

We have eliminated the sensuality that would otherwise allow young persons simply to move on to the next stage of development, always already an aspect of the young person's life as a whole movement into the future, thereby assuring the young person's own competence rather than that of the teacher alone. Bentham was well aware that the inspection principle invests authority in a person who takes up the inspector's or master's point of view. In this regard, he worried "whether the result of this high-wrought contrivance might not be constructing machines under the similitude of men," though even if we were to "call them machines: so they were but happy ones, I should not care":

> One thing only I will add, which is, that whoever sets up an inspection-school upon the tiptop of the principle, had need to be very sure of the master; for the boy's body is not more the child of his father's, than his mind will be of the master's mind; with no other difference than what there is between *command* on one side and *subjection* on the other. (64)

And again later he claimed that the inspection principle is not in itself "perverse" (66). Obviously Bentham had accepted notions of competence, liberty, and authority consistent with treating students as very young children, who, it must have been thought, *needed* to be inspected. Is it perverse to extend this need throughout everyday life?

At this point, then, we must consider more closely the notions of liberty and authority behind Bentham's acceptance of the inspection principle. To the extent that we adopt a cell-like posture throughout everyday life, we will take our liberty to be that of a person in a cell under inspection. Writing in 1843, Marx was still able to criticize this idea of liberty. In *The Declaration of the Rights of Man and of the Citizen* (1791), *The Declaration of the Rights of Man* (1793), *The Constitution of Pennsylvania,* and *The Constitution of New Hampshire,* the liberty in question, Marx concluded, is that "of egoistic man, man separated from other men and from the community":

> Liberty is thus the right to do and perform anything that does not harm others. The limits within which each can act *without harming* others is dete-

> mined by law just as the boundary between two fields is marked by a stake. This is the liberty of man viewed as an isolated monad, withdrawn into himself. . . . liberty as a right of man is not based on the association of man with man but rather on the separation of man from man. It is the *right* of this separation, the right of the *limited* individual limited to himself. (1967: 235)

For example:

> The right of property is . . . the right to enjoy and dispose of one's possessions as one wills, without regard for other men and independently of society. It is the right of self-interest. This individual freedom and its application as well constitutes the basis of civil society. It lets every man find in other men not the *realization* but rather the *limitation* of his own freedom. It proclaims above all the right of man "to enjoy and dispose of his goods, his revenues, the fruits of his labor and of his industry *as he wills.*" (236)

Given Bentham's idea that a prisoner is "indulged with perfect liberty within the space allotted to him" (47), we cannot help but draw the analogy to the liberty Marx criticized: my life-cell, so to speak, is bounded by the life-cells of the other members of my society, and as long as I do not cross these boundaries I am "indulged with perfect liberty." The "right of man" is that of a person who adopts the cell-like posture throughout everyday life, eliminating the sensuality that would otherwise allow the *realization* of one's own freedom through other persons.

The authority invested in an inspector is simply "political":

> The *constitution* of the *political state* and the dissolution of civil society into independent *individuals*—whose relation is *law* just as the relation of estates and guilds [in feudal society] was *privilege*—is accomplished in *one and the same act*. . . . The *egoistic* man is the *passive* and *given* result of the dissolved society, an object of *immediate certainty* and thus a *natural* object. The *political revolution* dissolves civil life into its constituent elements without *revolutionizing* these elements themselves and subjecting them to criticism. It regards civil society—the realm of needs, labor, private interests, and private right—as the *basis of its existence,* as a *presupposition* needing no ground, and thus as its *natural basis*. Finally, man as a member of civil society is regarded as *authentic* man, *man* as distinct from *citizen,* since he is man in his sensuous, individual, and *most intimate* existence while *political* man is only the abstract and artificial man, man as *allegorical, moral* person. Actual man is recognized only in the form of an *egoistic* individual, *authentic* man, only in the form of *abstract citizen*. (Marx, 1967: 240)

That is to say, authentic persons are the very persons whose "own powers" are "social powers" that must be realized *through each other,* as what Marx called "species-beings" (241) (and what we called "co-persons" in section one of this part). Were we to begin with such persons as natural persons, we would not need to abstract them as citizens. But in the "political revolution" we begin with egoistic persons as natural persons, hence the need to abstract authentic persons as citizens. In this regard Marx quoted Jean-Jacques Rousseau, who "correctly depicted the abstraction of the political man" (240), the citizen of the political state:

> Whoever dares [claimed Rousseau] to undertake the founding of a nation must feel himself capable of *changing,* so to speak, *human nature* and *transforming* each individual who is in himself a completed but isolated whole, into a *part* of something greater than himself from which he somehow derives his life and existence, substituting a *limited* and *moral existence* for physical and independent existence. *Man* must be deprived of *his own powers* and given alień powers which he cannot use without the aid of others. (*Social Contract,* Bk. II, London, 1782: 67) (241, Marx's emphasis)

Again, for Rousseau social powers are *alien* powers—not one's own powers—whereas for Marx just the opposite is the case:

> Only when the actual, individual man has taken back into himself the abstract citizen and in his everyday life, his individual work, and his individual relationships has become a *species-being,* only when he has recognized and organized his own powers as *social* powers so that social force is no longer separated from him as *political power,* only then is human emancipation complete. (241)

But if we live everyday life in a cell-like posture we will not subject "political power" to criticism, as did neither Bentham or Rousseau. We will be open to the very subjection that Bentham feared could come about only through a certain type of ruthless inspector. Here he misunderstood his own "engine," no doubt because he, like Rousseau, did not understand the basis of human nature in sensuality.

In many ways Marx echoed Musashi. He claimed that "the *senses* of social man *differ* from those of the unsocial"—or again, from those of Rousseau's "actual man"—inasmuch as in the latter case "*all* the physical and spiritual senses [including will, love, etc.] have been replaced by the simple alienation of them *all,* the sense of *having,*" the sense of egoistic man (1967: 308, 309). In a truly human emancipation, our senses will be social exactly as our powers will be social: "His own sense perception only exists as human sense perception for himself through the *other* man" (312). Mustn't we conclude that Marx aimed to celebrate a variety of *natural* sense perception? *The abstract citizen is invisible.*

The "sense of having" presupposes a certain reading of the world in which the object of sense stands apart at once as one's own. Meant exclusively, mine/not-mine is an ambiguity in the figure of an object of sense; neither side of it is visible. Hence *private property is invisible.* (*Actual* possession is nine-tenths of the law, as we say. Here we should recall what Vico discovered about "the ensigns . . . used as signs of the first division of the fields"; "in the time of the mute nations the great need answered by the ensigns was that for certainty of ownership" (1984: §486–7). *Capitalism constitutes a certain reading of the world.*) Still on their feet, American Indians had what William Cronon called "usufruct rights," with respect to which their using a thing was an aspect of natural movement that included other uses by other people. Their sense perception and the associated sense of freedom were "through the *other* man," that is, *social.* They did not need an "alien power" to *make* them social; they were always already social. Natural seeing and natural authority fit perfectly with Marx's notion of species-being (though, as we will develop in the following section, this notion must also include the co-making of space and viewpoints).

Political authority aims precisely not to allow a person to develop perception, understanding, and freedom "through the *other* man." One's own movements may not be efficacious directly through others. They may be efficacious through others only if they have been accorded that role, and that role is an aspect of a reading of one's movements, the guideline for which is *law,* established only at the abstract level of citizen. (To take just one example: a man and a woman may not have offspring directly through each other; the offspring will be illegitimate.) Hence, the perfect fit between political authority and the inspection principle: political authority is simply invested in inspectors at every crucial juncture of everyday life. Inspectors exercise what Foucault (1980) called "the eye of power."

We can get so used to this arrangement, moreover, that we take life without sensuality to be natural. We *assume* inspection. We no longer remember what Donzelot called "aleatory spaces," in which we could be *social,* "such as the vacant lot and the street, where sexual and amorous initiations were carried out" (229). (Note that we tend to avoid these spaces today, as the initiations have turned to the negative side of sexuality.) Aleatory spaces owed their special character to a life that remained sufficiently sensual, a life that continued for the lower classes long after it had been abandoned by the middle classes and the aristocracy. We once "rubbed shoulders" in a "mixing" of "the greatest possible variety of ages and classes," but then it "was all as if a rigid, polymorphous social body had broken up and had been replaced by a host of little societies, the families, and by a few massive groups, the classes" (Aries, 1962: 414).

It is one thing to seek individuality against the background of *social* life (Marx) or of *sociality* (Aries), but it is quite a different thing altogether to *begin* with an isolating individuality that may never be actually overcome, except, of course, in virtue of an alien, unnatural power. The inspection principle was pushed so far into our lives that we became egoistic even before we had a chance to subject the category to criticism: we became egoistic *in childhood.* Childhood was instituted to induce young persons to adopt cell-like postures.

How does a young person become a child? One is sentenced from birth to be a child. No overt act by a young person will earn that person release from the sentence to be a child, not even if the act can be read as deserving of such release. Until a young person comes of age, that person is disfranchised, legally a ward of its parents or equivalent adults. (Even thereafter a person does not automatically escape childhood, inasmuch as to be declared mentally incompetent in a civil proceeding is tantamount to returning to childhood (Scheff).) The same act that abstracts the adult, franchised citizen abstracts the disfranchised child. Child/adult is an ambiguity in the figure of our bodies; neither is visible.

It took about two or three centuries gradually to construct the institution of childhood. Only at the dawn of the twentieth century, for example, did we institute compulsory schooling. It is the hallmark of a culture of "descending individuality" that "the child is more individualized than the adult" (Foucault, 1979: 193). Here, as both Aries (1962: 333) and Foucault have pointed out, *discipline* is the key:

> In a disciplinary regime . . . individualization is "descending": as power becomes more anonymous and more functional, those on whom it is exercised tend to be more strongly individualized; it is exercised by surveillance rather than ceremonies, by observation rather than commemorative accounts, by comparative measures that have the "norm" as reference rather than genealogies giving ancestors as points of reference; by "gaps" [the features, the measurements, the gaps, the "marks" that characterize him and make him a "case"] rather than by deeds. (Foucault, 1979: 193)

Yes, especially not by deeds, any acts that are directly efficacious. Bentham had picked up the new sensibility for parents of his day, a sensibility that would gradually extend to parents at all levels of society, though it began in the upper echelons, and that would gradually be applied to all ages of young persons through adolescence, though it began with those short of puberty:

> This evolution corresponded to the pedagogues' desire for moral severity, to a concern to isolate youth from the corrupt world of adults, a determination to train it to resist adult temptations. But it also corresponded to a desire on the part of the parents to watch more closely over their children, to stay nearer to them, to avoid abandoning them even temporarily to the care of another family [as had been quite common]. The substitution of school for

> apprenticeship likewise reflects a rapprochement between parents and children, between the concept of the family and the concept of childhood, which had hitherto been distinct. The family centered itself on the child. (Aries, 1962: 369)

In *sociality* the family opens onto the "world of adults," shared by parents and their offspring alike as the arena of everyday life. But gradually the family turned inwards, onto itself, "centered on the child." We became "atomized individuals" (Aries, 1982: 613), especially because of, according to Donzelot, the "turning back [rabattement] of each of the members of the working-class family onto the others . . . at the cost of a loss of its coextensiveness with the social field," and "being isolated, it was now exposed to the surveillance of its deviations from the norm" (45).

On Bentham's idea of what parents want, young persons cannot be too much under the eye of the master, even when they retire at night and are not studying (which, as Aries pointed out, had a great deal to do with the emergence of boarding schools). But young persons are under a parent's eye even before they come to school. We have to be careful here because we can easily misunderstand what it is that parents do when they, to continue a manner of speaking, eye their offspring. Eyeing a young person can be an order of natural seeing or sensuality; the parents and their offspring are aspects of one whole movement, in which case we could easily imagine parents acting as Dionysius did with his "ear," spying on young persons to detect their infidelities. But, as Bentham knew so well, such eyeing does not induce young persons to feel as if they were *always* being seen.

Is a young person, even today, really confined to a cell under inspection? Well, recall that as an evening proceeds, all young persons are supposed to retire to their own beds. Can we imagine a point of view away from which all such beds stand apart at once? What movements by young persons directly overcome the separation of the beds in which they sleep? What movements by young persons directly overcome their lack of privacy in the beds in which they sleep? How many young persons, even when they are still in college but clearly of age, can return to their parents' home and lock the door of the room in which they sleep so that no one save themselves can enter?

Try to recall the trepidation of young persons who are keeping their first secret from their parents. What is the source of the trepidation? No matter where young persons are, don't they tend to refer their own acts to their parents' faces? They cannot help but conceive, as Bentham would have had it, that they are under inspection: the apparent omnipresence of their parents. When young persons cannot imagine a movement that eludes the gaze of their parents—or the gaze of surrogate parents, such as police or teachers—they must be in cell-like postures. They do not realize their freedom *through* their parents, but rather their parents are the *limitation* of their freedom. As *children* young persons are *read* by their parents: their being good or bad is an ambiguity in the figure of their acts that can be resolved only away from their parents' point of view.

By *law* parents do have the right of omnipresence; as children young persons do not have the right of direct action. If they try directly to overcome their separation from others in bed, their parents can even call the strong arm of penal justice, the police, to compel them to sleep alone. It is this network of all parents through a system of penal justice that constitutes the point of view away from which young persons are simultaneously isolated from each other. Is it not all too easy, as we did with schools, to imagine a layout of houses in a community, in which each house is broken down into rooms, some with beds, each labelled with the name of the young person who sleeps there? What a far cry from the times of Bentham, when so many homes must still have been like they all were not so long before: "It is easy to imagine the promiscuity which reigned in these rooms where nobody could be alone" (Aries, 1962: 395). It is a great feat, a part of the luxury of the institution of childhood, literally to realize the layout of houses and beds Bentham might well have imagined himself.

Aries made it clear that his reference to promiscuity was not a mere matter of imagination:

> This lack of reserve with regard to children [everything was permitted in their presence; they had heard everything and seen everything] surprises us: we raise our eyebrows at the outspoken talk but even more to the bold gestures, the physical contacts, about which it is easy to imagine what a modern psycho-analyst would say. The psycho-analyst would be wrong. The attitude to sex, and doubtless sex itself, varies according to environment, and consequently according to period and mentality. Nowadays the physical contacts described by Heroard [Henri IV's physician] would strike us as bordering on sexual perversion and nobody would dare to indulge in them publicly. This was not the case at the beginning of the seventeenth century. (1962: 103)

But by the beginning of the next century "the modern idea of childhood" had taken hold:

> In the regulations for the children at Port Royal we have Jaqueline Pascal writing: "A close watch must be kept on the children, and they must never be left alone anywhere, whether they are ill or in good health." But "this constant supervision should be exercised gently and with a certain trustfulness calculated to make them think that one loves them, and that it is only to enjoy their company that one is with them. This makes them love this supervision rather than fear it." This principle was absolutely universal. . . . The object was to avoid the promiscuity of the colleges. (114–5)

Here we are talking, it must be stressed, of young persons before puberty, for in society at large, if not in the colleges, young persons after puberty tended still to share "the world of adults." Even this limitation of the institution of childhood would end in the next century, the nineteenth, when we invented "the concept of adolescence" (268).

Now, Aries' claim about psychoanalysts is important for us to heed. At the dawn of the institution of childhood, the appropriate sexual epithet for young persons changed from immodesty to innocence. Young persons before puberty were once thought to be "unaware of or indifferent to sex"; one could hardly "soil childish innocence," as "nobody thought that this innocence really existed" (106). Then at the dawn of the twentieth century this innocence itself was exchanged for a childhood sexuality that was anything but innocent, at least in unconscious fantasy, if not in conscious reality as well. As the family closed in on itself, to occupy what we may call "a family cell"—among the inspectors of which were, crucially, psychoanalysts—the young person came increasingly to be seen as needing to navigate very tricky sexual waters *in the family itself* in order to turn up at puberty and thereafter as a so-called normal adult (Donzelot).

We once were born merely into an order of sensuality that extended beyond family boundaries. Foucault and Sennett have even shown that, in the times in which we still lived largely in an order of sensuality, sexuality was thought of as *between* persons, *naturally social,* as Marx might have put it. Yet by the dawn of the twentieth century Freud would take sexuality to be *in* persons: exactly as a century earlier we began to think of each person as having his or her own freedom, we began to think of each person as having his or her own sexuality, not to be realized through other persons.

Psychoanalytic theory is a certain reading of our bodies, and the only qualified inspectors are psychoanalysts. Conscious/unconscious is an ambiguity in the figure of our bodies; neither is visible. Just to entertain the ambiguity we have to take our bodies to occupy readable space. In keeping with his times and training, Freud went so far as to imagine that a mechanical biology constituted the ultimate foundation for his resolution of the ambiguity. But, no matter how we had imagined the constitution of body in readable space, we would not thereby have been in a position to perform a psychoanalytic reading. How did we come to this impasse in the first place? We had to institute childhood and a family centered on children in such a way that the simultaneous isolation of the members from each other, and of the family from every other family, could displace the original order of sensuality—an order of *joint* movement—in which we all had once lived. *Sexuality as we know it is based on the distortion of sensuality by the cell-like posture.*

It was not enough to sweep ourselves off our feet into the space of the world in view, essentially separate from each other, indeed, not enough even to do so as if our bodies were in cells under inspection, at least not until the inspectors were psychoanalysts. As we extended childhood into a young person's life, we gradually constructed inspection-sexuality. At first inspection aimed only to isolate young persons from any person who might have introduced him or her into adult sexuality; as in the panopticon itself, young persons were left at least with the liberty of their own cells. But then we extended childhood past puberty. By the dawn of the twentieth century—and Freud played no small part—we had invented the reading of "masturbatory insanity," acccording to which young persons' attempts to realize themselves by touch "at a more mature age"—Freud's own terms—were *bad,* symptoms of *illness,* to be resolved only away from a psychoanalyst's face (Szasz, 197). So much for openings to realize anyone by touch! Only contact of the limits of our bodies remained.

Again, it was not enough to distort sensuality by the cell-like posture, turning our sensuality back on ourselves so as to induce us to have our own sexuality apart from others; even that sexuality had to come along with ambiguities that could be resolved only away from a psychoanalyst's face. So long as we returned to our feet when it came to sexuality we could not be effectively confined to a family-cell. We have to ask ourselves today, not only who regularly realizes us by touch when we are young persons, either before or after puberty, but who regularly realizes us by touch after we come of age. Sexuality as we know it tends to displace *all* of our sensuality. (To take just one example: same-sex friends are certainly not generally sensual, mostly out of fear of being sexual and thereby earning another *bad mark,* as Foucault might have put it.) Yet to be *social* we must be *friends on our feet, realizable by touch throughout our everyday lives.*

Today we can, as we say, have sex, even lustful sex, without making love. The sex, we go on to say, is merely physical—of the body we leave behind in space—whereas the lovemaking is more than physical—of the soul we invent to leave our bodies behind in space. It is certainly no wonder that we still admire Plato's, rather than Freud's, analysis of love. But what we do not notice is that even our lovemaking is not sensual. We often complain that after making love we still do not feel touched; off our feet, we may make contact only with the limits of each other's bodies, as if we were playing the other's body to make music for the soul, though music always open to falling on deaf ears. Hence, sex therapists often recommend that we *tell* our lovers how to play the right music with our bodies. We do not together-step here anymore than we do elsewhere.

Sometimes we do feel realized by touch, but these times just happen to us: we *fall in* and *fall out* of love. When we fall in love with another person, it is as if that person were inserted into our very bodies; even when the other person is not in our presence we feel drawn out of our bodies toward that person. We often think of this feeling as being a mysteriously magical feeling, inasmuch as, once we are swept off our feet, such sensuality is not a matter of choice—*not an option*— let alone clearly understandable. We also say, for example, that we can undress each other with our eyes, but in doing so we make nudity the resolution of an ambiguity in our appearance, especially true—or rather, oppressively true—of woman's appearance. If, on the other hand, we undress each other on our feet, nudity is simply visible, though we must move to realize it. (Recall here our remarks about clothes in part one above.) We have eliminated the sensuality in which *any* two persons can be aspects of a whole movement that unites them in the realization of each other.

IV. Sensual Friendship

Today we do not have the faith that we can resolve our conflicts without referring them to an inspector's face. But then again we also do not have *the option* of together-stepping with each other, the solidarity of sensuality. *Trust is simply the solidarity of sensuality.* It constitutes one level of our common sense, that of the order of co-making of movement and standpoints. In the previous section, we outlined how this level gradually lost its significance in everyday life over the last two or three centuries. In this section, we will extend our discussion of the solidarity of sensuality, but we will also discuss the other level of our common sense, that of the co-making of space and viewpoints. The solidarity of sensuality is only a necessary, not a sufficient, condition for realizing our common sense.

When we were on our feet in the world, the solidarity of sensuality was a sufficient condition for realizing our common sense. If all separation is essentially a function of movement, then one's life may be a movement into the future. One grows up as an aspect of the whole movement of one's life and that of one's parents; one learns through others as a whole movement into the future. Displaying the common sense of sensuality, Black Elk's father began to defer to him even while he was young enough to be regarded as a child. The adult-child relationship in an order of sensuality is reversible, as indeed any relationship of natural authority must be reversible. Today the adult-child relationship is not only irreversible (Goffman, 1979), but also *not natural,* as Musashi might have put it.

The institution of childhood broke apart the process of life so that the life in which one is born is divided from the life in which one dies. Life is no longer a whole movement of which birth and death are but two aspects, both always already a part of life. (Think of the wheel of becoming in Buddhism. But also recall, following Aries, that only in the last century did we begin, indeed, were we even able, to keep death from the world of young persons. Not long enough ago for us to forget so completely, young persons themselves were much more likely to die than they are today, and moreover they were just as likely to live with only one parent because of their other parent's death as today they are deprived by divorce.) Whatever general stage of development was recognized in the whole movement of life, it had to be visible, only a movement away, allowing an extraordinary person such as Black Elk directly to move to it. Now, however, adulthood has become a reading of one's life, and the reading of adulthood—*the citizen*—is the basis of *political* authority.

We can apply these ideas equally well to the teacher-student relationship. Is what the teacher has only a movement away? Is the future state in which one knows always already an aspect of one's life as a whole movement into the future, thereby assuring one's own competence rather than that of the teacher alone? Learning is thereby a whole movement that involves both teacher and students as well. Or, on the contrary, is one's competence the resolution of an ambiguity away from the teacher's point of view so that it is not open to natural sense perception? If so, one is *immediately* in need of being inspected—the metamorphosis of needing to learn into needing to be taught and of needing to grow into needing to be raised—*as if inspection were natural rather than political.*

Again, one cannot tell by oneself that one has learned by being taught; one must be inspected by the teacher, and once one comes to assume inspection in this way, one is a student who must be taught, as opposed to a person who must learn. *Exactly as a child is invisible, so is a student.* In an order of sensuality, on the other hand, competence is simply this: *I can.* We do not have to read a young person; we can simple *see* one *move* in the required way. We need to acquire a sense of the very specifics of taking this result seriously. Fortunately we need not invent these specifics. In his experimental First Street School, as we already noted in the introduction, Dennison celebrated the idea that learning is not the result of teaching, but rather an aspect of a whole movement that involves both teachers and students. Allowing Dennison to speak as much as possible for himself, let us look more closely at First Street.

José, a thirteen-year-old boy, came to First Street with what his public school called "a reading problem." "To say 'reading problem,' " Dennison claimed (76–7), "is to draw a little circle around José . . . ignoring everything about José except his response to printed letters." In his typical way, Dennison understood reading as an order of movement, having "an organic bond" with speech:

> Printed words are an extension of speech. To read is to move outward toward the world by means of speech. Reading is conversing . . . [there is] an organic bond between reading and talking. (76–7, 165)

Hence, Dennison began his work with José by noting very carefully how José made a place in the world with his body as he approached a reading lesson:

> He had been talking animatedly in the hall. Now as he comes to join me, his face contracts spasmodically and the large gestures of his arms are reduced to almost nothing. There is no one near him, and he is absolutely free to refuse the lesson, yet he begins to squirm from side to side as if someone were leading him by the arm. He hitches up his pants, thrusts out his lower lip, and fixes his eyes on the floor. His forehead is lumpy and wrinkled like that of a man suffering physical pain. His eyes have glazed over. Suddenly he shakes himself, lifts his head, and squares his shoulders. But his eyes are still glassy. He yawns abruptly and throws himself into the chair beside me, sprawling on the tip of his spine. But now he turns to me and smiles his typical smile, an outrageous bluff, yet brave and attractive. "Okay, man—let's go." (78)

José was even able to read Spanish when he came to this country at the age of seven, and yet over the last five years in his public school, no doubt along with the rest of his environment, he had been induced into not being able to move outward toward the world by reading:

> He could not believe, for instance, that anything contained in books, or mentioned in classrooms, belonged by rights to himself, or even belonged to the world at large, as trees and lampposts belong quite simply to the world we all live in. He believed, on the contrary, that things dealt with in school belonged somehow to the school, or were administered by some far-reaching bureaucratic arm. There had been no indication that he could share in them, but rather that he would be measured against them and be found wanting. Nor did he believe that he was entitled to personal consideration, but felt rather that if he wanted to speak, either to a classmate or to a teacher, or wanted to stand up and move his arms and legs, or even wanted to urinate, he must do it more or less in defiance of authority. (80–1)

No whole movement into the future united him with his books and his teacher; he was not even as fortunate as early human beings who could vicariously refer the world to a face through a together-step of their faces with their leader's face. Without such a base on which to develop his own point of view, the only point of view that could be at work was that of his teacher.

Without the terms of posture, however, Dennison was driven to think of José's problem as a problem in *seeing:*

> When I used to sit beside José and watch him struggling with the printed words, I was always struck by the fact that he had such difficulty in even *seeing* them. I knew from medical reports that his eyes were all right. It was clear that his physical difficulties were the sign of a terrible conflict. On the one hand he did not *want* to see the words, did not want to focus his eyes on them, bend his head to them, and hold his head in place. On the other hand he wanted to learn to read again, and so he forced himself to perform these actions. But the conflict was visible. It was as if a barrier of smoked glass had been interposed between himself and the words: he moved his head here

> and there, squinted, widened his eyes, passed his hand across his forehead. The barrier, of course, consisted of the chronic emotions I have already mentioned: resentment, shame, self-contempt, etc. (81)

The barrier is easier to understand in our terms: José *can see* the printed words but *cannot view* them. Because of the way in which the book in his hands was always already referred to his teacher, José could not refer the book to his own face. When it came to reading, we can say, he did not really have his *own* face.

It is crucial to stress again, as we already did in the introduction in other terms, that Dennison did not try immediately to address José's face as if it were detached from the order of sensuality of his body. If "one single formula" constitutes the remedy, Dennison claimed, it will be "this physical formula: bring the bodies back":

> I wanted to get back, in some way, to the stage of reading at which written words still possessed the power of speech. And so our base of operations was our own relationship; and since José early came to trust me, I was able to do something which, simple as it may sound, was of the utmost importance: I made the real, the deeper base of our relationship a matter of physical contact. I could put my arm around his shoulders, or hold his arm, or sit close to him so that our bodies touched, or lean over the page so that our heads almost touched. Adults, and especially adult Americans, are not used to this kind of touching . . . [but] the importance of this contact to a child experiencing problems with reading can hardly be overestimated. (168–9)

Heeding the relationship between very young persons and their immediate others when they learn to speak in the first place, Dennison made the "deeper base" of his relationship with José *sensual* so that they could *realize each other by touch*. He aimed to encourage José's sense of direct efficacy in reading as in speaking, to *move with the book* toward the world. The point of view required to resolve the ambiguities in the printed words must come on the basis of this sense of efficacy:

> Nor was it my body beside his that meant so much, but the fact that the presence of my body vivified his awareness of his. He knew where he was: he was in his skin; and when the little bursts of panic made his head swim or his eyes turn glassy, he did not have to run away or reject the task *in toto*. He could gather himself together, because his real base—his body—was still there. (169)

In gathering himself together José could realize, in effect, that the "causes of learning" are in himself, not in the teacher:

> When we consider the powers of mind of a healthy eight-year-old—the avidity of the senses, the finesse and energy of observation, the effortless concen-

> tration, the vivacious memory—we realize immediately that these powers possess true magnitude in the general scale of things. Besides them, the subject matter of primary education can hardly be regarded as a difficult task. (73)

Should we take the causes of learning to be in the teacher, on the other hand, "learning is difficult" (74), and José's "reading problem" encircles him once again. Referring to the work of both Leo Tolstoy and Paul Goodman, Dennison argued that "learning to read is a far easier task than learning to speak" (190). Goodman himself concluded that, other things equal, any normal young person "will spontaneously learn to read by age nine, just as he learned to speak by age three," that is, "unless he is systematically interrupted and discouraged" (quoted in Dennison, 190).

Perhaps, as Dennison entertained, young persons can learn to read without the organic bond between reading and talking, but only if they can have their own faces to put in the light that reflects from the printed words. For the many reasons Dennison cited, José just did not have his own face when it came to reading. The simplest way to oppress a person in today's society, indeed, is to create a barrier to having one's own face *throughout* everyday life. Dennison also made it clear that José was precisely such a victim: José came to live in a "defiance of authority" that returned him to, as it were, the mass of faceless subjects, not "entitled to personal consideration." It is all too easy for us to forget that, as swept off our feet as we may be, it is still to our own bodies that we must refer the world. In the process of constructing inspection-schools, we relied on a residue of sensuality to provide the real basis for our young persons to learn to read.

Fortunately, the principle of inspection has not penetrated to the arena in which young persons learn to speak, no doubt because "few indeed will insist that it was they who taught their children to talk" (89). (A very young person who has learned to speak may well still conflate left/right and east/west, as any parent knows. Hence, the principle of inspection *cannot* penetrate here.) It is in learning to speak, claimed Dennison, that we find the best evidence for the body as the real base of persons:

> Two features of the growth of this mastery are striking:
> 1)The infant's use of gestures, facial expressions, and sounds is at every stage of his progress the true medium of his being-with-others. There is no point at which parents or other children fail to respond because the infant's mastery is incomplete. Nor do they respond as if it *were* complete. The infant, quite simply, is one of us, is of the world precisely as the person he already is. His ability to change and structure his own environment is minimal, but it is real: we take his needs and wishes seriously, and we take seriously his effect upon us. . . .

> 2)His experimental and self-delighting play with sounds—as when he is sitting alone on the floor, handling toys and babbling to himself—is never supervised and is rarely interfered with. . . . The play goes on . . . absolutely freely. . . . The infant, in short, is born into an already existing continuum. (90)

But we could easily overlook the physical base of this continuum if we think of the infant as we would think of ourselves in his place, that is, as imitating grown-up speech, a practice that would "remove the instrumentality he has all along been studying":

> Parents are perhaps deceived because the sight is inevitably so charming, all that display of participation, with so little content. They forget that among their own motives the desire to charm, to enliven, to make a merry noise, is not insignificant. And they overestimate the content of what they themselves have said, for the truth is that the music of our ordinary conversations is of equal importance with the words. It is a kind of touching: our eyes "touch," our facial expressions play back and forth, tones answer tones. We experience even the silence in a physical, structural way; they, too, are a species of contact. In short, the physical part of everyday speech is just as important as the "mental.". . . . The infant is surrounded by the life of the home, not by instructors or persons posing as models. Everything that he observes, every gesture, every word, is observed not only as action but as a truly instrumental form. And this indeed, this whole life of the form, is what he seeks to master. (91–2)

Dennison obviously advocated the original sense of thinking/speaking/reading: all three are ***between*** us in an order of sensuality in which we ***realize each other by touch***. (His use of the term 'contact' here and above is obviously not meant to refer to the order of contact of the limits of bodies.) Precisely "this whole life of the form" constitutes the real base of a person. A person is from birth embedded in this order, "actually taking part" in the on-going order of movement of speech, "a true social sharing in the degree to which he is capable":

> The infant, in short, is not imitating but doing. The doing is for real. It advances him into the world. It brings its own rewards in pleasure, attention, approval, and endless practical benefits. This very distinction between imitating and doing . . . is the root meaning of "learning by doing." (93)

Hence, we return to the direct efficacy that only an order of sensuality promotes. Once we remove young persons from "this whole life of the form" by placing them in an inspection-school, we break apart the order of sensuality in which they could have "advanced into the world." As Bentham knew so well, we induce them to assume their own inspection *even before* they try to advance into the world.

Unlike the reformers who first introduced the institution of childhood, therefore, Dennison must advocate an entirely opposite conception of "natural relationships," what we advocated in the introduction in terms of the relationships of persons to each other:

> I mean that when adults stand out of the way so children can develop among themselves the full riches of their natural relationships, their effect on one another is positively curative. Children's opportunities for doing this are appallingly rare. The school life is dominated by adults, and after school there is no place to go. The streets, again, are dominated by adults, and sometimes by juvenile violence which in itself is an expression of anxiety. . . . Perhaps the most important thing we offered the children at First Street was hours and hours of *un*supervised play. . . . Indeed, on several occasions with the older boys, I averted violence simply by stepping out of the gymnasium! . . . One might speak—without any sentimentality at all—of the *gay intelligence* of childhood. This is not to discount the passions and gravity of childhood, or its incidental violence, or the fitful opacity which sometimes makes children, like animals, seem mysterious; it is only to observe that once children are in motion among themselves, the quality of gay intelligence becomes apparent immediately and characterizes their games. (82–3, 212)

Moreover:

> I would like to quote Rousseau again and again:". . . do not save time, but lose it." If Susan had tried to save time by forbidding the interesting conversation about voodoo, she would first have had a stupid disciplinary problem on her hands, and second (if she succeeded in silencing the children) would have produced that smoldering, fretful resentment that closes the ears and glazes the eyes. How much better it is to meander a bit—or a good bit—letting the free play of minds, adult and child, take its own lively course. . . . The children will feel closer to the adults, more secure, more assured of concern and individual care. Too, their own self-interest will lead them into positive relations with the natural authority of adults, and this is much to be desired, for natural authority is a far cry from authority that is merely arbitrary. Its attributes are obvious: adults are larger, are experienced, possess more words, and have entered into prior arrangements among themselves. . . . These two things, taken together—the natural authority of adults and the needs of children—are the great reservoir of the organic structuring that comes into being when arbitrary rules of order are dispensed with. (24–5)

Yes, the *natural authority* of *persons* who are *visibly* older, larger, experienced, and so on, thereby making the authority only a movement away. Obviously,

Dennison depended on creating a solidarity of sensuality, a trust, entirely absent in inspection-schools.

Dennison's understanding of young persons undermines the institution of childhood:

> 1) we trusted that some true organic bond existed between the wishes of the children and their actual needs, and 2) we acceded to their wishes (though certainly not to all of them), and thus encouraged their childish desire to take on the qualities of decision-making. (21)

The young persons actually practiced what would be demanded of them as older persons, and this practice went as deep as possible into their education:

> So many adults these days live in a world of words—the half-real tale of newspapers, the half-real images of television—that they do not realize, it does not sink in, that compulsory attendance is not merely a law which somehow enforces itself, but is ultimately an act of force: a grown man, earning his living as a cop of some kind, puts his left hand and his right on the arm of some kid (usually a disturbed one) and takes him away to a prison for the young—Youth House [another inspection-facility that Foucault showed to be a crucial aspect of a disciplinary regime]. I am describing the fate of hundreds of confirmed truants. The existence of Youth House, and of the truant officer, was of hot concern to two of our boys. They understood very well the meaning of compulsory attendance, and understanding it, they had not attended. We abolished the act of force, and these chronic truants could hardly be driven from the school. (88)

But it is a mistake to think here that Dennison was either giving freedom where it did not already exist, or acknowledging the type of freedom that Marx had criticized so well. Freedom, for Dennison, is a matter of *wholeness,* which is well illustrated precisely in the matter of sexuality:

> Aside from regret for the past, and the anxious fear of regret itself, the adults were not speaking (either pro or con) of the actual behavior of young persons. They were speaking of their own willingness to sanction sexuality, or their desire to confine it. Nor were the young who were present worried very much that their own pleasures would be denied them. They would outwit the adults, as always. They would climb out windows and down rainspouts; they would tell lies, etc. . . . Sanction is not a matter of *what happens,* but of methods of control and ideals of life: what sort of world must we build? The young understand this question with great immediacy. And they want a great deal more than sexual freedom. They want *wholeness.* They do not want to lie and evade and suffer guilt, but to affirm themselves in the largest possible harmony of self and society, passion and intellect, duty and pleasure. . . . The problem begins in infancy and runs through the whole society. Mere slogans and attitudes are of little value, and there is a terrible,

> naive hubris in the behavior of a teacher who believes that he can sanction sexuality by conferring freedom. He can no more sanction it than sanction the law of gravity. Beyond this he can ally himself with the student's quest for wholeness. Here the teacher's own quest for wholeness is extremely valuable. . . . In life, as in art, the healing truth is the whole truth. The libertarian teacher cannot give freedom. He can only cease to control. He cannot sanction sexuality. He can only seek to allay guilt. (102–4)

Today our young persons must fall back on whatever remains of the order of sensuality that would otherwise be entirely open to them were we to cease to isolate them from each other according to the principle of inspection. Sensuality, again, would otherwise envelop all of us in its solidarity and trust, allowing students and teachers to seek wholeness *together.*

Although the illustration in sexual matters is crucial, especially due to the original justification for the institution of childhood, it must also be stressed that, our typical attitudes notwithstanding, the free play of sensuality that forms the real basis for learning is not simply a sexual matter. Dennison was not alone in recognizing the special nature of children "in motion among themselves." In Colin Ward's defense of play (88), he quoted at some length from *The Peckham Experiment,* which we also quoted in the introduction:

> The boy who swings from rope to horse, leaping back again to the swinging rope, is learning by his eyes, muscles, joints and by every sense organ he has, to judge, to estimate, to *know.* The other twenty-nine boys and girls in the gymnasium are all as active as he, some of them in his immediate vicinity. But as he swings he does not *avoid.* He swings *where there is space*—a very important distinction—and in doing so he threads his way among his twenty-nine fellows. Using all his facilities, he is aware of the total situation in that gymnasium—of his own swinging and of his fellows' actions. He does not shout to the others to stop, to wait or move from him—not that there is silence, for running conversations across the hall are kept up as he speeds through the air.
>
> But this "education" in the live use of all his senses can only come if his twenty-nine fellows are also free and active. If the room were cleared and twenty-nine boys sat at the side silent while he swung, we should in effect be saying to him—to his legs, body, eyes—"You give all your attention to swinging; we'll keep the rest of the world away"—in fact—"Be as egotistical as you like." By so reducing the diversity in the environment we should be preventing his learning to apprehend and to move in a complex situation [or rather, in an aleatory situation]. We should in effect be saying—"Only this and this do; you can't be expected to do more." Is it any wonder that he comes to behave as though it is all he *can* do? By the existing methods of teaching we are in fact inducing the child's *inco-ordination* in society. (Pearse and Crocker, 192)

If the boy swings to *avoid,* he adopts an egoist liberty; the others are the limits of his own freedom. Instead, in *natural* relationship with others, he swings *where there is space;* he realizes his own freedom *through* the others, a *coordinated* movement of which all involved persons are aspects together, *in sociality.* What an eloquent appeal to us to understand the importance of the whole movement of all of us, not only the young among us! How awkward and shy so many of us tend to feel in the complex situation of the world of bodies in which we live, and yet our sensuality, so evident in us before we assume our own inspection, is itself only a movement away!

Now, in an order of sensuality, as Foucault put it (1979: 49), "by breaking the law, the offender has touched the very person of the prince." So far in part three we have discussed the gradual loss of sensuality as if our being "touched"—for better or for worse—were to go along with it, and with it alone. Dennison commended the "kind of touching" we do in ordinary conversation. Is this "kind of touching" merely an order of sensuality?

In the introduction, Goffman's work on "social organization" (1972) helped us to understand, albeit in other terms, that the answer must be negative. He studied "insanity of place" and the special nature of "mental symptoms" because, once we understand how things go wrong in "social organization," we can also understand how they go right:

> It is important now to emphasize that a social deviation can hardly be reckoned apart from the relationships and organizational memberships of the offender and offended, since there is hardly a social act that in itself is not appropriate or at least excusable in some social context. The delusions of a private can be the rights of a general; the obscene invitations of a man to a strange girl can be the spicy endearments of a husband to his wife; the wariness of a paranoid is the warranted practice of thousands of undercover agents. Mental symptoms, then, are neither something in themselves nor what is so labeled; mental symptoms are acts by an individual which openly proclaim to others that he must have assumptions about himself which the relevant bit of social organization can neither allow him nor do much about. It follows that if the patient persists in his symptomatic behavior, then he must create organizational havoc and havoc in the minds of members. (1972: 356)

And again:

> In ceasing to know the sick person, they cease to be sure of themselves. In ceasing to be sure of him and of themselves, they can even cease to be sure of their way of knowing. A deep bewilderment results. . . . And life is said to become like a bad dream—for there is no place in possible realities for what is occurring. It is here that mental symptoms deviate from other devia-

> tions . . . [in which the] grammaticality of the activity is sustained. . . . [A mental patient's] behavior strikes at the syntax of conduct, deranging the usual agreement between posture and place, between expression and position. (366–7)

Goffman's use of some of the terms of posture was by no means an accident. Once we leave our bodies behind in space, we enter an order of co-making of points of view. We must read each other *through each other:*

> But the fact is that there is no stable way for the family to conceive of a life in which a member conducts himself insanely. The heated scramble occurring around the ill person is something that the family will be instantly ready to forget; the viable way things once were is something that the family will always be ready to re-anticipate. For if an intellectual place could be made for the ill behavior, it would not be ill behavior. It is as if perception can only form and follow where there is social organization; it is as if the experience of disorganization can be felt but not retained. . . . Further, as already suggested, there are a multitude of reasons why someone who is not mentally ill at all, but who finds he can neither leave an organization nor basically alter it, might introduce exactly the same trouble as is caused by patients. (382, 387)

Even in the order of the co-making of the space of the world in view, we can say, by breaking the rules of social organization, the offender "touches" the very person of the offended. As R. D. Laing also argued (Laing, 1970; Laing and Esterson, 1970), a person can actually be induced to develop "insanity of place" by being accorded, through the activity of his or her immediate others, what Goffman would have regarded as not being a "*workable* definition of himself" (366). José's behavior when confronted with the task of reading can easily be read to constitute "insanity of place" in the normal inspection-school classroom. Without Dennison's sensitivity, very few teachers indeed would believe that it was that very classroom that had induced José's behavior in the first place.

Once we organize ourselves according to the principle of inspection, the "touching" of co-making does not have enough presence in our everyday lives. By managing, at least in some cases, to give this "touching" enough presence, Dennison showed us just how crucial it is. If we organize ourselves in an order of sensuality as much as we can do, how do we also organize ourselves in the space of the world in view so that the "touching" of co-making will have enough presence in our everyday lives to open the way to realizing our common sense?

The social anarchist Giovanni Baldelli advocated a practice by which we gradually eliminate *law.* One by one we eliminate specific laws, and in the process gradually develop our skills at living without law. We could easily transform this practice into one that will gradually return us to a life in sensuality, or at least in as much sensuality as we can have. One by one we eliminate acts of inspection, and in the process gradually develop our skills at living without inspection. Or again, one by one we allow movements to be *directly efficacious* in life. This practice can be engaged at home, school, and work, and Dennison's First Street School is a good example of it.

At the same time, as Dennison also realized in his own way, the practice of removing acts of inspection affects the orders of both viewing and seeing: *to engage in the practice of removing acts of inspection is to aim to be a person,* at least in the sense of that concept we developed in section one of this part. But because we cannot give up the special abilities that came along with getting off our feet, we must *realize each other* through viewing exactly as we realize each other through seeing. Off our feet we can still realize ourselves as persons to the extent that *co-making is evident to us:* we are *co-persons,* realized through each other, in an order of co-making. Moreover, we need not invent a practice to make our being co-persons evident in everyday life. We can simply practice *the principle of consensus.*

On application of the principle of inspection, good/bad is not a matter of consensus, neither the automatic consensus of sensuality, in which case good/bad is visible, nor the negotiated consensus of various points of view. We believe that the practice of *inspection-politics* is our only way out of chaos—to urge the practice of consensus today, for example, is to risk "insanity of place"—yet we have also organized everyday life so as to exclude sensuality. We are left with a sense of individual isolation and impotence, exactly the condition that promotes the political authority of inspectors in the first place. Which came first: the failure of faith in consensus, or the elimination of sensuality in favor of surveillance, inducing us to repress our ability to togetherstep with each other to accomplish our desired goals?

On application of the principle of consensus, points of view are regarded as *equally valuable precisely as the points of view of persons:*

> Consensus is based on the belief that each person has some part of the truth and no one has all of it, no matter how we would like to believe so, and on respect for all persons involved in the decision that is being considered. . . . The assumption is that we are all trustworthy (or at least can become so). (Estes, 25)

Each and every point of view offers a *necessary* ingredient in the process of negotiation. By practicing negotiation to consensus one learns that not to live up to this standard of equality is simply not to arrive at the "correct" decision:

> This is always one of those times when feelings can run high, and it is important for the meeting, or group, not to use pressure on those who differ. It

> is hard enough to feel that you are stopping the group from going forward, without having additional pressure exerted to go against your examined reasons and deeply felt understandings. In my personal experience of living with the consensus process full-time for 12 years, I need to say that I have seen the meeting held from going forward on only a handful of occasions, and in each case the person was correct—and the group would have made a mistake by moving forward. (Estes, 28)

But exactly this standard of equality is excluded by the principle of inspection. As we will develop in more detail in the final section below, if progress cannot be only a movement away, based simply on direct action, that is, if the task is to resolve an ambiguity in the world, each point of view should be granted the status of one among others in *a circle of consensus*—a circle both because a circle cannot stand apart at once, facing from its beginning to its end, and because, to the extent that a circle can stand apart, each participant makes an equal place. (Again, in the terms of posture, it is best to understand this requirement *literally.*) Each participant may then be *a person,* embracing the *common* sense of *co-persons.*

Now, none of what has just been said should be taken to imply that at other levels of description, persons are equally valuable. In the order of sensuality, for example, persons are equally valuable to the extent that they share the way their faces are referred to the world. Each may still have unique abilities, as indeed Black Elk did have. And again, these abilities are visible: *I can.* We simply never take these various and sundry abilities to indicate that we are not equally valuable at the base of our lives in sensuality, in the co-making of movement and standpoints. Hence, for example, decision-making is dispersed in classless societies.

As well, when off our feet, each of us may still have unique abilities to read the world. If we do not turn any of these abilities into the practice of inspection, we will realize our equal value at the base of our lives off our feet, in the co-making of space and viewpoints. We will be, again, co-persons. Hence, we will *realize through each other* that our unique abilities are aspects of an order of co-making. *This is human emancipation* in the sense in which Marx intended it, for our individual powers are realized as *social* powers: the common sense of co-persons. Social organization will be *evidently* social; we will feel, literally feel, that our powers are realized only through each other.

Dennison's First Street School was an extended experiment in the common sense of co-persons. He rightly took the first step to be one of sensuality, building the solidarity required to form circles of consensus. The young persons could actually exercise their natural authority to contribute to the course of their own learning. Although the First Street School failed—and here Dennison's honesty was most helpful, as well as most eloquent—it did so because all of the participants were still embedded in a society based on the principle of inspection. Were that society itself, even in the most gradual fashion, to reinforce experiments like the First Street School by the practice of being a person—again, by removing acts of inspection one by one—such experiments would meet with increasing success.

What, to begin with, should we try to do in the family? As parents we now tend to stand apart from our offspring to inspect them. We imagine as well that we can keep secrets from them in a way in which they cannot keep secrets from us. For example, we may issue vague warnings along with rules that aim to exclude certain movements: "you cannot engage in this or that activity, at this time or at that place." Such a practice is notorious for inducing just the movements that we were aiming to exclude. Moreover, the so-called disappearance of childhood is frequently lamented today because the available media present these forbidden movements to young persons in ways that make them symbolic of adulthood (Postman). (As we will discuss again in the final section below, by confusing the need to learn to read with the need to be taught to read, Neil Postman failed to recognize the *political* nature of the institution of childhood, both at its origin and in its subsequent history.) Perhaps individually we can keep the required secrets, but otherwise little hope remains.

So, bring the bodies back, as Dennison put it. *Make the basis of family life visible.* Establish the order of sensuality of one's family, working with each other *exactly as Dennison worked with José.* Each person must feel that his or her movements may be directly efficacious. Look for every way in which to promote this, especially on the part of young persons. In such an order of sensuality, enough trust exists to realize that young persons will seek natural authority just in case they do not know what they are doing, and otherwise will learn the hard way, which is precisely the way of learning most commonly cited by older persons when they make exclusion rules in the first place. Or again, in such an order of sensuality, enough trust exists to realize that the practice of protecting young persons for their own good actually induces them not to realize, as Dennison put it, the "true organic bond" between their wishes and their actual needs, that is, their own natural authority. (This kind of trust is crucial, as we now know, even for very young persons; if we do not trust our offspring *right from the very beginning* we end up as inspectors who attempt to read the moment when they are ready to be trusted, and thereby we will fail to ally ourselves with their "quest for wholeness." To make daycare work, for example, we must recognize it as just another occasion for their growth and learning. We can exercise care in choosing daycare without being inspectors ourselves, and indeed without looking for surrogate inspectors.) To issue vague warnings along with exclusion rules is simply to push young persons away, possibly to get into trouble in hiding, just where they cannot make use of anyone else's natural authority either.

Many young people who live in inspection-families fail to notice that they present themselves to their parents in the hope of having their own readings of themselves confirmed by their parents, even in the hope of establishing friendships with their parents, but that their parents do not reciprocate—and indeed parents cannot reciprocate if they wish to maintain their status as inspectors. I suggest to these young people that they try what I began to call "the osmosis method" once I realized the significance of sensuality: "Get on your feet, in a purely sensual relationship with your parents, and do not get off your feet again until your parents give up being inspectors and present themselves as if they were in a circle of consensus." On the osmosis method, one must resist presenting a reading of oneself, or of others and the world, especially if it is in conflict with the reading given by one's parents. (Note that one need not also abandon the behavior, or rather the movement, about which the conflict of readings has arisen.) One must express whatever feelings one has for one's parents that can be expressed by realizing them through movement, such as a hug, a smile, a remark about the weather, as we say—always keep in mind what Dennison called "the music of our ordinary conversations"—or through the lack of movement, such as simple silence—always keep in mind the force of an appropriately timed visit to the bathroom! Simply put: if it cannot be said by osmosis, it should not be said at all.

The osmosis method can be applied in any asymmetrical relationship, even by a parent of young persons who fail to present themselves as if they were in a circle of consensus. Indeed, because the method is founded on the co-making of movement, it can be applied in *any* relationship. (Some of the young people to whom I have recommended the method have eventually applied it in the context of their employment, both as employers/managers and as employees, and thereby caught a glimpse of the ground on which the organization of work can be *between persons*. In a more playful vein, we can note that in such a "naturalistic setting" as a "holiday beach" "postural congruence is a very real, common phenomenon . . . that operates at a very low level of consciousness," perhaps "by visual clues in peripheral vision" (Beattie and Beattie)—or rather, more generally, by a kind of osmosis, that of the co-making of movement.) No matter what the relationship, as soon as participants *try* to apply the osmosis method they appreciate its force: it immediately breaks the spell of readings, a spell under which we take the conflict of readings to constitute our entire relationship with the other participants.

When, however, it comes to resolving ambiguities in the world once the solidarity of sensuality is reestablished (or established in the first place), rules for reading may well need to be set. Again, not every pattern of behavior follows from sensuality. (Here I am thinking of something as simple as exactly when to go to sleep. To the extent that we read time, we do not go to sleep merely in terms of how our faces are referred to the world, though we should do so as much as possible.) Whatever rules for reading are to be set, let them be set through circles of consensus, in the open. Parents, for example, must try to resist *arbitrarily* breaking a circle by saying "because I said so"—and to resist is surprisingly easy once bodies are brought back—for otherwise the principle of inspection would be reestablished, with the rules of inspection a secret once again. Parents will hereby begin to realize the way in which a family is *between* its members, the basis on which the natural reasons for or against engaging in this or that activity, at this time or at that place, will simply come into the open, undermining the notorious family feuds over how to read the world.

The practice of removing "child/adult" and replacing it with "person" is just as important to aim at the adult. Not only do parents tend to stand apart from their offspring, but they also tend to stand apart from each other. We know enough now about the nature of the level of invisibility in our lives to know that man/woman, husband/wife, and father/mother tend to be readings, the resolutions of ambiguities in the figures of our bodies, not visible at all. So, again we must ask ourselves what exactly is visible in older persons. What remains of these three ambiguities if we return first to an order of sensuality, and then complement it with circles of consensus? We will discover the persons at the base of each of us, allowing our unique abilities without at the same time breaking us apart.

For example, at what level of description do we decide who works for money in the family? The person who is visibly a woman may find it difficult to defend her working at any level other than that of a person. As a woman, wife, or mother—to the extent that these are invisible—she tends to be read as not as well-suited for working as the person who is visibly a man; he can find easy justification for working at all three levels of his invisibility, as a man, husband, or father. Hence, to the extent that we live simply as persons, this disparity must disappear altogether.

To take another example, at what level of description do we decide who initiates sexuality between lovers, say, between a person who is visibly a man and a person who is visibly a woman? And at what level of description do we decide how far lovemaking will go between such lovers? We know how we now tend to read the answers to these questions, especially how we read our bodies to resolve intercourse as the ultimate end of lovemaking. (As we noted in part one, remember, the Maoris once spoke of the moon as "the true husband of all women.") It is no accident that the most effective sex therapy is an application of the osmosis method, namely, to make love without intercourse; it induces us, wittingly or not, to be *persons* who make love by realizing each other through *a whole movement* of our bodies. (Intercourse should be reintroduced once it is not motivated by a reading, once it is entirely an aspect of the whole movement of our bodies again.) To the extent that we live simply as persons, we not only undercut the present-day dilemmas of invisibility that trap so many of us in unsatisfying relationships; we also provide the ground on which *any* two persons may be lovers.

Indeed, *if off our feet we make socialization a process of consensus,* the only remaining disparities could not be divisive in the way they are now, in the competition of standing apart. *We would simply be co-persons* with each other as well as with our offspring. Through our common sense we would recognize our unique abilities and then put them to use in our attempts to survive as families in this world. The same could be said about a married couple without offspring, or of two or more people who choose to live together out of love or friendship. Whatever the unit of immediate living with each other, the practice of being persons opens the way to realizing our common sense.

Obviously, however, this practice must reach beyond units of immediate living, which now stand apart from each other in isolation. Common sense will reestablish (or establish in the first place) the extended families, neighborhoods, and local communities that were broken apart in our rush to push the principle of inspection as far as possible into our everyday lives. And as the solidarity of sensuality is gradually realized, we will no longer believe that the sheer number of people in a circle is an obstacle to consensus. We will find a way to move with respect to each other that will eventually allow a consensus to emerge. No doubt this order of movement will include an organization of subdivisions of people not unlike some of those that exist today (Ward), though the overall difference of evident co-making will make all the difference in the world.

For example, the institution of childhood isolates learning from working, but not as a matter of movement itself. No movement can directly overcome the separation. We need to ask ourselves, as we gradually remove acts of inspection, how to make the boundary between learning and working a matter of movement again. When and where do each of the activities of learning and working take place? If we spread both over the duration of life as well as over the arena of life, we may well blur the boundary between them, but not yet eliminate it altogether: activities of learning and working will come together in different kinds of *enabling centers*. (Again, the institution of childhood—not to learn by doing—is a luxury that must also be recognized as ecologically unsound, as a form of pollution, an inefficient use of energy. To learn by doing kills two birds with one stone, as we say.) As we will discuss in more detail in the final section below, visible differences between us and between enabling centers can still remain, and in cases in which differences of reading—invisible differences—must arise, we can work them out in circles of consensus. The advantage is that, no matter who one is or where one is, both learning and working will be only a movement away.

As we will also discuss in more detail in the final section below, to return to being a person is also naturally empowering with respect to the practice of inspection-politics. How can we have some sort of impact even as inhabitants of a state of inspection-politics? *By moving together!* (Even voting is most effective as a together-step.) Few of us today realize this means of reestablishing control over our own lives because we feel isolated and impotent in general. But, inasmuch as co-persons do not read each other as citizens of a state of inspection-politics, the practice of being persons cannot help but induce bewilderment in that state, opening the way to our natural authority to together-step to realize our desired goals. Much has already been written about our ability to take up this challenge (Ward). It suffices in this section to point out that this tradition in anarchist literature is supported by the simple practice of being persons and realizing our common sense.

As we gradually return to visibility and consensus of viewpoints, we will begin to enjoy *sensual friendship,* a natural extension of the original friendship at the foundation of, for example, the Ho'o Pono Pono we discussed in part one, section four. Instead of the myriad of ways in which we now read each other into invisibility, we will simply be sensual friends: visible as much as possible, and otherwise at work *in the open* on consensus of viewpoints. This practice will certainly blur every boundary that we now read, though thereby it will undermine all feuds over how to read the world: *we will let loose the natural healing energies of our common sense.*

V. The Co-Making of Inquiry

We began our quest for the origin of inquiry by studying the difference between a bee's orientation in the world and a modern human being's orientation in the world. Very early in our history we were like bees, as our faces were automatically referred to the world, and then we gradually achieved what I have called "loose faces," faces to which the world is referred. This achievement marked the origin of inquiry as we know it today. Although early human beings engaged in a kind of inquiry, it was on their feet in the visual world, where the interrogation of the world must be conducted in what Vico called "natural language," at its extreme a language that Black Elk thought he shared with other inhabitants of the world (Neihardt). We, on the other hand, encounter problems just in sharing our native languages with other human beings, let alone with the other inhabitants of the world.

We came to have a life in which we referred the world to our own faces. The term 'own' is crucial here: each of us came to have his or her own face in a way that early human beings could never have known. In the introduction and throughout this part, I have commended the ability to return to the level at which we do not have our own faces and can practice what I called "aleatory inquiry." Given our seduction by the posture of making places at points of view—and, as I remarked in the first section of this part, it is the seamless quality of attention deflection that makes this seduction possible—we tend to assume that returning to the level of aleatory body/person entails abandoning the sense of self that makes inquiry possible. But far from constituting the chaotic state that we imagine it constituting from our own points of view, aleatory inquiry establishes trust: the solidarity of sensuality. Sensuality is not the antithesis of inquiry, but rather the ground of inquiry as co-persons, the order of co-making of inquiry.

Moreover, throughout the book, especially in part two, I have argued that we do not even understand the nature of the inquiry we do conduct. Vico, of course, would have taken this question as to the nature of inquiry as a question about its origin, or rather, as he would have preferred to say, as a question about its making. Today we do not understand that this making of inquiry is a co-making. Nor do we understand the extent to which inquiry still depends on sensuality. I have tried to illustrate our lack of understanding by using Dennison's First Street School as an example of a context of inquiry—an actual arrangement of students and teachers, along with their props—that runs counter to what we typically understand the context of inquiry to be. I then expanded Dennison's method into what I called "the osmosis method," which returns us to our feet as much as possible in our everyday lives. I recommended that to resolve problems in living we begin with the osmosis method and then, on the basis of the solidarity of sensuality it makes possible, form circles of consensus to negotiate our individual points of view.

Now, when it comes to illustrating an alternative arrangement of people around those who possess special knowledge at a higher level in our lives than that of the First Street School, I still use the osmosis method, this time to create the example of an aleatory university, which is analogous to the example of an aleatory street I used in the introduction. Let us suppose that every day each student must rediscover what readings apply to the buildings and rooms as well as to the people he or she will encounter that day. It is a rare occurrence indeed that a student is not provided with a map that determines how to read buildings and rooms—readings often set in concrete, as it were, on the very buildings and rooms—so as to discover the whereabouts of specific people, especially the persons read as "experts" who are supposed to possess the knowledge the student wishes to acquire. The use of the term 'possess,' as Dennison showed in the case of José, must not be taken lightly here: a "discursive regime" is in sway. Typically university students are not really in a different posture from that of José. They too are taken to be in need of being taught rather than merely in need of learning and, as it were, must ask permission to go to the bathroom unless they wish to run the risk of being inappropriate in a way that could certainly damage their careers. But without a map that determines how to read buildings and rooms, university students would need to begin their learning at a more fundamental level than they ever are allowed to begin today.

To some extent the level would be purely aleatory, as they move to be near first one, then another, gathering of people in various rooms. "What," they will need to ask, "is going on here?" A lot would need to be learned just to begin to answer this question. Were the gatherings of people in circles of consensus, moreover, the inquiring student could move quite smoothly from the purely aleatory level to the level of the negotiation of points of view, the level at which in a typical university today he or she would be subject to the "discursive regime" of "experts." But again, what if the "experts" were always in circles of consensus in which their points of view were no longer "eyes of power," precisely as the points of view of persons. Then the ensuing negotiation would constitute a perfect setting for the inquiring student to participate in the interrogation necessary to real inquiry. The concept of Dennison's First Street School would be extended to the university itself.

Well, of course, we could not have universities that called upon students to begin *every* day at the purely aleatory level, though an aleatory day or two would prove extremely valuable to all the participants, including those who possess special knowledge. The example of an aleatory university is useful in establishing an image of an alternative context of learning in which persons read as "students" literally stand on equal footing with persons read as "experts." The way one can move becomes crucial to the arrangement of persons at work in the university. But most crucial is that everyone goes beyond visibility and the solidarity of sensuality only in circles of consensus, which could just as well take place in *any* context, no matter how aleatory it commences.

I want now to return to the result of part two in order to remind us of the ground for this proposal about the conduct of communities of inquiry. The aim is ultimately to reopen the dialogue begun in the introduction about the nature of inquiry. There we spoke of Galileo and Kuhn to introduce a certain problem of objectivity that we already resolved in part two. So far in this part we have concentrated on the co-making of inquiry in the context of early education, though we have at least sketched the extension into everyday life. We need to concentrate here on the so-called leading edge of research, the level of inquiry that we call upon to resolve our most difficult problems in thinking and living.

To begin the review, then, the basic lesson to be drawn from the special theory is that light is the ordering *net* of physics. What were formerly independently located reference bodies in space are now dependently located in space-time; or rather, it is by reference to the *movement* of light *between* them that they play their roles as aspects of *events* in space-time. Observers, along with their clocks and rulers, are just as caught up in this net of events. As

Einstein put the classical version, "the simultaniety of the two definite events with reference to one inertial system involves the simultaneity of these events in reference to all inertial systems"; "this is what is meant when we say that the time of classical mechanics is absolute" (149). For the special theory, on the other hand, "the sum total of events which are simultaneous with a selected event exist, it is true, in relation to a particular inertial system, but no longer independently of the choice of the inertial system"; " 'now' loses for the spatially extended world its objective meaning" (149). Unlike classical observers, special-theory observers are crucial to the particular way in which events are ordered. Without relating observers to each other in terms of the *movement* of light *between* them, we cannot even begin physics or indeed any inquiry that requires a temporal ordering. When, for example, sociologists approach persons in interaction, mustn't they be prepared to determine a temporal ordering too: just what is the order of events in society? Although we do not tend to appreciate this level of inquiry in sociology as we do in physics, we cannot avoid engaging in it, even if we do so only tacitly.

To understand the significance of this level in general, let us reconsider Einstein's original problem, but now with *two* observers in spatial frames of reference:

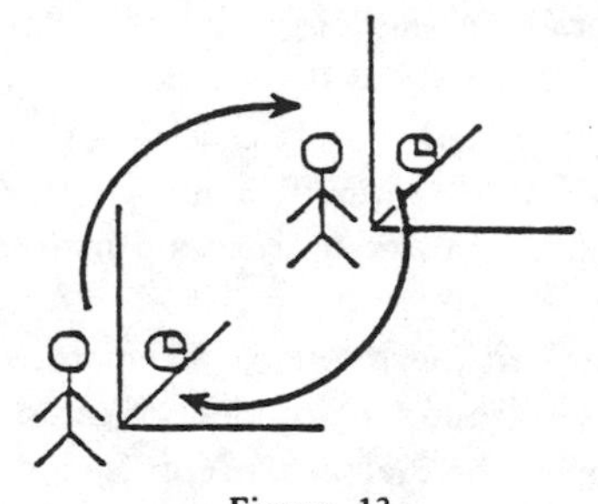

Figure 13

We cannot help but see here two persons in interaction—*a social relationship*—in which they attempt, so to speak, to synchronize the terms they employ to communicate about their respective clocks. Each person may well be using his or her own temporal terms: until they do synchronize their temporal terms, they cannot interpret each other's reports, and until they do interpret each other's reports, they cannot synchronize their temporal terms.

Although Einstein was properly concerned only with the "logical circle" of temporal terms, we may just as properly be concerned with the "logical circle" of *any* terms. How do we *begin* to interact with each other, precisely when we do not as yet know that our terms are synchronized: until we do synchronize our terms, we cannot interpret each other's reports, and until we do interpret each other's reports, we cannot synchronize our terms? We run in a "logical circle" at the origin of all of our inquiries, even within our own society or culture. Hence, to begin inquiry at all, not only in temporal terms, we must make a series of free choices, to define our terms, as we say. Hereby we *make* the invariance that grounds our communications and eliminates the possibility of mere relativism.

To give a rough illustration of this resolution of the problem of communication, let us imagine two observers who report to each other about the ostensive grounds of their respective terms by pointing to aspects of their worlds and at the same time writing the term they employ to refer to that aspect, say, as its label. Let us imagine further that the performances of observers one and two are observed at the midpoint between them in exactly the way that Einstein required for temporal synchronization:

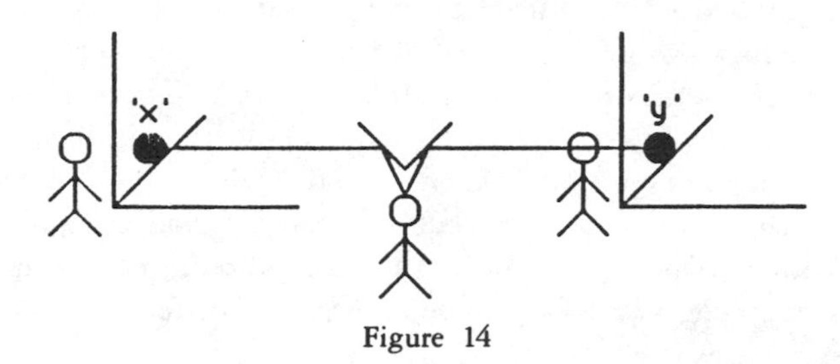

Figure 14

Why not simply say, following Einstein, that the terms of observer one and the terms of observer two are synchronized just in case they appear that way to observer three at the midpoint between them? Hereby we are consistent with the special theory and at the same time we provide a procedure for the synchronization of all other terms as well, though as we move away from ostensive grounds—and perhaps even there as well—we must no doubt imagine much more complex feats of mirror synchronization, if indeed mirror synchronization will still suffice.

If we understand ostensive grounds on the level of our faces being referred to the world, then mirror synchronization must suffice: *it reflects precisely the solidarity of sensuality.* As we already noted just above and at the beginning of part two, here we work at the level Vico would have called "natural language." Originally inquiry began *between* persons because they were visible by default. They did not even need mirrors. Nor would we need them today were our language entirely a "natural language." But as soon as we use, so to speak, unnatural or arbitrary signs in the place of natural ones, signs which are referred to our own faces, mirrors may well still suffice, but only if that to which the signs refer is *entirely in the visual world.* We will then know that, regardless of its social or cultural context, each human face is referred to the world in essentially the same way. Again, as we noted at the beginning of part two, it is at this level that we talk about the sensitivity of the human eyes to color (Lukes, 267). Whatever the differences between societies entirely on their feet in the visual world, these differences must be resolved by light alone.

W. V. O. Quine spoke of "the pattern of chromatic irradiation of the eye" and the associated concept of the "affirmative stimulus meaning" of a sentence: "the class of all the stimulations that would prompt assent" to the sentence (31–3). He then went on to speak of cases of sentences that are "independent of stimulus meaning" (52), cases in which we cannot tell the difference between two sentences on the basis of the irradiation of the eyes alone. Here, obviously, the faces in question must no longer be referred to the world, for otherwise the difference between the two sentences could be resolved by light alone. It is crucial to realize that Quine's most well-known case—the one in which he asks, "Who knows but what the objects to which this term applies are not rabbits after all, but mere stages, or brief temporal segments, of rabbits?" (51)—vanishes completely in the visual world, where the difference in question is no difference at all. Quine realized in effect that, if aspects of the world away from our faces are referred to our own faces, mirror synchronization will not suffice (unless it is confined to terms that can be said to come along *with* a face whose orientation is referred to itself—these terms must have, as it were, purely ostensive recipes).

It is crucial to note again that in the case of the synchronization of temporal terms the velocity of light for one-way trips is *stipulated.* The velocity is not resolved by light alone, but rather must be referred to our own faces: it is stipulated by reference to an observer's face placed at the midpoint between two clocks. Indeed, Einstein realized that this placement is a free choice. In this case too mirror synchronization does not suffice *to tell* the "stimulus meaning" of temporal terms such as that of 'simultaneity.' Or again, irradiation of the eyes of the third observer does not determine simultaneity. The first two observers must agree to abide by the judgment of a third observer placed at the midpoint between them, and then they can interpret each other's reports. To begin physical inquiry, in other words, *the observers must enter into a circle of consensus.*

Einstein's case is exactly like Quine's well-known case, in that the differences between the possible definitions of simultaneity vanish completely in the visual world. The relativity of simultaneity with respect to spatial frames of reference gives way to the invariance of intervals in space-time. The first two observers in Einstein's case must be referring the world to their own faces. Were they Hopi, the problem of interpreting each other's temporal reports would be resolved by default: they would both be on their feet, *at standpoints,* in space-time, without experience of the "now" that loses its objective meaning for the spatially extended world. Or again, were the first two observers working with two, different definitions of simultaneity—that is, with two, different stipulations about the velocity of light for one-way trips—this difference would make no difference at all to a third observer at a standpoint. Hence, the members of the community of physical inquiry that concerned Einstein had been swept off their feet into the space of the world in view, and were in need of entering into a circle of consensus in order to begin their inquiry.

What is not clear in Einstein's work itself, in other words, is that his terms are unnatural or arbitrary, referred to our own faces, in a way that eventually leads us into quantum mechanics, as noted in part two, and at the same time into circles of consensus, as noted just above. Einstein's free choice is both *explicitly social* and *a reading—hence, a co-reading: inquiry is a co-making, whether on or off our feet.* The choice of the velocity of light is really *between* observers, as pictured in the diagram above. Or more generally, only *between* persons do we find the ground on which their terms may be synchronized: *all inquiry begins there.* Einstein's own work now becomes just another element in what can be regarded as a general program for the synchronization of the terms we use to communicate with each other, a general program that establishes, as also noted in the first section of this part, the common origin of sociology and physics.

To put our result in its most general form we can draw the analogy between Wittgenstein's analysis of the meaning of terms and Einstein's analysis of the meaning of terms. Wittgenstein's rejection both of a private language and of what he called in his early work "the picture theory of language"—that language reaches out to touch the world (1972:15)—bears an interesting resemblance to Einstein's rejection of the classical significance of simultaneity in one inertial frame. Another way to grasp the significance of the claim that inquiry begins between us is to realize, following Dennison again, that learning a language in the first place is just such an inquiry.

Wittgenstein likened the problem of meaning in a private language to the dilemma of a person who consults another copy of today's newspaper to check on the accuracy of the first copy's reports (1968:§265). If I have a certain experience which I define to be of X, then later I may have another experience which I believe to be of X. But if I try to check the accuracy of my belief by recalling my original experience of X, that is, if I try to recall what 'X' means, this recollection is just another experience that also may or may not be of X, as if I were, on analogy, checking another copy of today's newspaper.

If 'X' is to have meaning in a private language, in other words, the person in question must follow a rule that connects one experience of X to another. But what could constitute following a rule in this case?

> If our considerations so far are correct, the answer is that, if one person is considered in isolation, the notion of a rule as guiding the person who adopts it can have *no* substantive content. There are, we have seen, no truth conditions or facts in virtue of which it can be the case that he accords with his past intentions or not. As long as we regard him as following a rule "privately," so that we pay attention to *his* justification conditions alone, all we can say is that he is licensed to follow the rule as it strikes him. This is why Wittgenstein says, "To think one is obeying a rule is not to obey a rule. Hence it is not possible to obey a rule 'privately'; otherwise thinking one was obeying a rule would be the same thing as obeying it" ([1968:]§202). (Kripke, 89)

Confined to my own experience in this way, I have trouble quite like I have in physics when I am confined to my own inertial frame. In the latter case, I have no ground on which to claim that two signals of light that arrive simultaneously at my place began that way—or alternatively, that two events which seem to me to happen simultaneously happened that way—whereas in the former case I have no ground on which to claim that a present experience that seems to me to be qualitatively identical to my original experience of X is actually identical to it. Indeed, I can no more attach meaning to the object of an original experience than I can attach meaning to the date of an original event. In both cases, I try to reach out from the present to the past but cannot tell whether or not I have succeeded.

Put another way, the analogy here is between Wittgenstein's rejection of language as reaching out to touch the world and Einstein's rejection of our ability to *measure* the velocity of light for a one-way trip (or even to measure the difference in the velocity of light for round trips of equal distance in different directions). If we can measure the velocity of light, we do not need to get between observers to tell whether or not their temporal terms are synchronized. Or alternatively, if we can touch the world with our language, we do not need to get between language users to tell whether or not their terms are synchronized. If I am not confined to my own experience or to my own inertial frame, I do not so much succeed in telling whether or not I have reached from the present to the past as I agree to enter into a certain community of fellow language users: my fellows and I must reach a consensus as to how we will exercise free choice at the origin of our langauge.

To believe that we can determine the relationships between terms and the world—that language reaches out to touch the world—is to believe that our terms are not so much related to each other as they are to the world.

What will hold for one user of terms will hold for all others. We will have a classical theory of absolute meaning—the old sense of objectivity—rather than a special theory of the relativity of meaning, what in general Wittgenstein thought of in terms of "language games," games whose significance is derived from the relationships *between* the players (1968:§69ff).

> The rough conditional thus expresses a restriction on the community's game of attributing to one of its members the grasping of a concept: if the individual in question no longer conforms to what the community would do in these circumstances, the community can no longer attribute the concept to him. Even though, when we play this game and attribute concepts to individuals, we depict no special "state" of their minds [or alternatively, no special way in which they refer the world to their own faces], we do something of importance. We take them provisionally into the community, as long as further deviant behavior does not exclude them. In practice, such deviant behavior rarely occurs. (Kripke, 95)

Although the problem of understanding deviant behavior will have to be reconsidered below, Saul Kripe's understanding of the roots of language games is entirely apt: *the roots must be circles of consensus.*

Now, it must also be stressed that these roots run deeply into the world as well. The new sense of objectivity developed in part two extends the concept of community as it was understood in philosophy by Wittgenstein and Kripke, indeed, as it was also understood by the "new sociologists of science" (Restivo) we referred to in the introduction. If the issue of the "social construction of reality" is raised as if the problem of "an independent reality" were still to be resolved one way or another, or as if it were still possible to "transform epistemology from a philosophical to a sociological project" (Restivo), then a variation on the old sense of objectivity is still in sway, now with the stress on the side of the human community as opposed to the side of the world. Sal Restivo claimed both that "an alternative science should not be conceived in terms of alternative scientific laws or techniques but rather in terms of alternative institutions and societies," and also that "some sort of objective knowledge is possible." Again, if the old sense is possible, with or without a stress on the side of the human community, then we fail to bridge the gap between ourselves and the world, perhaps still "an independent reality." And if the new sense is possible, "epistemology" does become "a sociological project" of inquiry, but no more so than a "philosophical" or a physical one. The transformations that Restivo advocated are automatic: *all inquiry begins between us and the world.*

As soon as we consider "alternative institutions and societies," we must consider a "new politics of truth" that displaces the current "discursive regime," as we put it in the introduction by citing Foucault. We have finally reached the point at which we can elaborate this project. It is crucial here to understand exactly how the circles of consensus of everyday life merge with the circles of consensus of the leading edge of inquiry. In the first section of this part, we developed the concept of a person and the associated concept of a uniquely human good—a good referred to our own faces—that is an aspect of an order of co-making between co-persons. The current "discursive regime" maintains its hold on the uniquely human good by obscuring both its own ground in circles of consensus and the more general ground of the circles of consensus of the whole society. Hence, the ground on which we could trust each other to remake our society is not evident to us.

In both the introduction and the last section, the so-called reading problem of José served as a crucial example of the impact of the current "discursive regime" on the young persons of our society. Although Foucault also would have analyzed this example in terms of the inspection principle, the special contribution of the terms of posture is to enable us to realize the very invisibility of the governance in question, the governance that Marx would have thought of as "political authority." What Bentham could not achieve by a mechanical contrivance in the visual world is automatically granted to the authority—the "eye of power"—of José's teacher. José's lack of any sense that he could somehow possess the words he was being asked to read must be placed in a context in which, so to speak, José could not have put his finger on the source of his problem. It is another matter altogether, a matter in the visual world indeed, should someone have literally taken his book.

The transformation of the need to learn into the need to be taught is the crux of the institution of childhood. Postman argued that this institution was required by the new demands of learning to read in a world that had come to turn on the written word. No doubt new needs to learn did arise in our history, especially, as Postman noted, after the invention of the printing press, but what is not clear is the need to place this learning in a context of teaching which is organized, as it currently is, by the inspection principle. Once organized as such, teaching entails the invisible governance that induced José's illiteracy: he became paralyzed by a sense of his own inadequacy to do the required work. Goodman's contention that a normal young person will simply learn to read soon enough if he or she grows up in an appropriate learning environment is the antithesis of Postman's argument: we can at least consider arranging ourselves in an alternative way so that we accomplish what Postman thought could be accomplished only through a transformation that obscured the efficacy of "natural authority" (Dennison).

The need for an alternative arrangement is more pressing than we typically realize today because we are so extensively organized by the inspection principle. The reading "teacher" serves as a paradigm for the reading of all sorts of inspectors of our lives, and hence as a paradigm for the pervasive presence of invisible governance. The subtle way in which this paradigm has come to function in our lives needs to be reinterpreted in the terms of posture so that we will understand how to resist what Thomas Scheff would have thought of in terms of our "suggestibility" to be "labelled" according to the paradigm. Not only our very bodies have come to be read so as to refer automatically to the "eye of power" of an inspector. So, too, have the props that we throw along with ourselves, as I put it in the introduction, in the betting game of our lives today.

The most general way in which the level of readings comes to entail an invisible governance is constituted by our sense of "insanity of place." If our perception "can only form and follow where there is social organization" (Goffman, 1972), we cannot avoid the ground on which invisible governance enters our lives: we can feel out of place in a way that threatens our very sense of self. Once we came to work so extensively in terms of readings in social organization, we were "suggestible" to a new type of governance, and some of us were quick to take advantage of it. Now we would be hard pressed to remember what life was like without it. As we noted in section three, at each crucial juncture of social organization some person is read as having not only a special expertise to deal with the associated breakdown of social organization but also an "eye of power": our very sense of expertise involves invisible governance. The reading of "expert" entails an authority to perform the readings of "good/bad" that allow/do-not-allow passage from one status to another in society. When we feel out of place, in what terms do we automatically think?

Ivan Illich, Thomas Szasz, Goffman, Laing, and Scheff have all testified to the crucial role that the terms used in the mental health profession play in our lives. Here I am thinking of the professional therapists who are sanctioned to exercise their skills by the state—the paradigm of which is a psychiatrist—so that their work does not involve the "natural authority" simply to "understand persons" (Laing, 1971a), but rather the "political authority" to "label" persons as "competent/incompetent," that is, as fit/unfit to be accorded the rights of a "citizen" of the state (Scheff; Szasz). We use the label 'sick' quite often in everyday life, typically when someone steps out of "place" in social organization. But it is one thing to mean "out of formation," quite another to mean "off course" (Laing, 1970), and if the readings of "good/bad" we use in everyday life involve the latter meaning, then the terms of the mental health profession have transformed the need to be well into the need to be healed or cured (Illich). Sick/well, in this sense, is an ambiguity in a person's body resolved only away from the relevant professional's face. Hence, if we tend to read this face *into each other's* behavior, we subject ourselves to the invisible governance of the mental health profession, exactly what Scheff meant when he spoke of our "suggestibility." Hereby, the control of a very general passage from one status to another in our society is taken out of our hands, and no more crucial passage exists as far as our rights as citizens in our society are concerned (Scheff).

After the resurgence of the feminist movement some twenty years ago, feminists documented another subtle way in which this sort of control of passage shapes our lives. I am thinking here of the reading of a person as "woman." At that time, mental health professionals tended to think of the paradigm of adult psychological well-being as essentially equivalent to the paradigm of male psychological well-being, and hence a person read as "woman" came to be thought of as lacking a certain psychological well-being (Chesler, 1972; 1973). (A mental health professional would work in terms of a paradigm of well-being that called upon the person read as "woman" to adjust to her condition in society, for example, to a marriage understood on the paradigm of male psychological well-being (Bart; Bernard; Steil).) As we also noted in part one, Carol Gilligan subsequently expanded this research to establish the existence of *two* paradigms of psychological *well*-being—in her terms, *two* images of self and relationship—one of which tends to be found in persons read as "women" and the other in persons read as "men." In her summary of Gilligan's research, Sandra Harding put it this way:

> The feminine "relational" emphasis on developing increasingly complex and intense forms of cooperation and intimacy in human relationships, and on developing the empathy and sensitivity necessary for taking the role of the particular other (the particular person to whom one is relating), is systematically regarded as less mature than the characteristic masculine "objectifying" emphasis on developing increasingly complex forms of achieving separation from others and on learning to take the role of the generalized other (anyone who stands in this kind of relationship). (52)

If one adheres primarily to the paradigm that tends to be found in persons read as "men," one's image of maturity is thereby directed away from the alternative paradigm of psychological *well*-being, and yet this paradigm must be understood as a derivative of the new sense of objectivity developed in part two.

"Is gender," Harding asked, "a variable in conceptions of rationality?" Again along the lines of Gilligan's research, especially on moral reasoning, Harding sketched two paradigms of rationality—so to speak, *two* paradigms of philosophical *well*-being—and on the one for "men," in contrast to the one for "women," "a rational person values highly his ability to separate himself from others and to make decisions independent of what others think—to develop 'autonomy'" (53). But the paradigm found in persons read as "men" is not found in all such persons:

> The fact that not only women but also Afro-Americans and working-class Americans are recorded on scales such as Kohlberg's to have less than a "fully human" sense of ethics, and that this kind of conception of the highest "human" morality is so consistent with the conception found in the writings of modern moral thinkers such as Kant and Rawls, should remind us that we are examining here not masculine and feminine rationality per se, but particular modern forms of gendered rationality. Why there is such a "coincidence" between the modern ideal of a citizen and the ideal masculine personality in our society is yet to be fully explained. (Harding, 55)

Without doubt, however, if the persons read as "experts" in these matters adhere to the paradigm of moral and rational development found in certain persons read as "men"—let us say, "the male paradigm of inquiry"—not only do most persons immediately appear to be less than adequate, but the old sense of objectivity is thereby reinforced: inquiry must aim to develop the one correct reading of a person, "the generalized other." This reinforcement is also circular, of course, because it is precisely the old sense of objectivity that leads us to think in terms of autonomy with respect to each other *and to nature.*

As Harding also noted (48), the work of Evelyn Keller and Carolyn Merchant established most clearly how the old sense of objectivity was all along a male paradigm of inquiry: "science itself . . . appears to practically everybody today (Keller) and at its origins (Merchant) to be even more closely tied to characteristics thought to be disproportionately distributed to men than are the making of war, health, laws, governments, and art"—the institutions in which certain persons read as "men" hold sway. The male emphasis on one correct point of view has always served to obscure the tendency to weigh one point of view over another in a hierarchy of points of view, as if the male point of view were *the* point of view anyway. It does not even matter, however, that the hierarchy of points of view favors males over females in particular. If the paradigm of inquiry is essentially that of the inspection principle, then the object of inquiry—which was nature herself in the inquiries that Keller and Merchant studied—must be reduced to the lower echelons, those subject to inspection in the first place: hence nature was separated from us exactly as we were separated from each other. Merchant often referred to the work of Francis Bacon, who surely took himself to be inspecting nature as if nature were "a common harlot" in desperate need of measures of control (171). Although the metaphor that guides the work of current inquirers is certainly not Bacon's in spirit, it is still in effect just as demeaning of the object of inquiry, as we have noted above in the case of mental health professionals: the current "nonperson" treatment of "clients" by medical "experts" (Goffman, 1961) was originally extended to nature, thereby constituting what Merchant called "the death of nature," and today many of us have begun to fear that this death will be literal.

It is not the case, however, that either the new sense of objectivity or "the female paradigm of inquiry" fails to applaud the reading of nature. They simply take all points of view to constitute *mutual* relationships *with* each other and the rest of nature. The terms of posture alone allow us to realize that ultimately these relationships must be grounded in the solidarity of sensuality so that they can be formed into the circles of consensus necessary to the co-making of inquiry itself, as we will continue to argue below. Whenever expertise is not visible, it may still flourish in circles of consensus: one point of view of nature may well have something special to contribute here or there even if it has no other weight attached to it in a circle of consensus, especially not the circle-breaking weight of an "eye of power."

So, again, the case of reading a person as "woman" is an example of the subtle way in which a person's behavior or movement can be referred to a certain point of view—in this case, the point of view of "man"—thereby constituting that point of view as an "eye of power." What is otherwise entirely personal is immediately in an institutional context that is organized according to the inspection principle. The resulting problem is profoundly troubling in just the way that Restivo believed that Kuhn should have been worried about the institution of science: the discussion of the community structure of science must realize the way in which readings of the world are socially constructed (207, 217).

Any institution that applauds the old sense of objectivity can also relieve itself of responsibility for the demeaning judgment involved in its readings of "clients." Not only can an "expert" adopt a special point of view with inspector's privileges, but the object of this point of view can also be thought of as an independent reality that allows only one correct reading. Inasmuch as *this* reading of the world is independent of the community structure of the institution in question, any judgments rendered thereby will still seem as if they were not socially constructed—hence, the new sociology of objectivity that aims to establish the social construction of *all* readings of the world (Restivo). Or again, in upholding a standard of mental health that is grounded according to the old sense of objectivity, a mental health professional can avoid being accused of merely upholding a dominant ethic (Szasz). (Prior to the version of the old sense of objectivity that constituted the advent of modern science, we upheld a religious version, the residue of which we still find in Galileo's thinking about the way God wrote "that great book," "the universe"—hence, the inquisition, during which the persecution of persons read as "witches" was conducted as if its readings were anything but socially constructed.) Szasz posed a most profound question about our ability to overcome the tendency to "scapegoat" at the foundations of our societies, that is, the tendency to establish persons read as "good" at the expense of those persons read as "bad" (287).

At the advent of modern science, indeed, we scapegoated nature, in the name of women to boot (Merchant).

The old sense of objectivity, even with the stress placed on the side of the human community as opposed to the world, will always manage to lead us into inquiries as if we were face to face with some sort of antagonist in need of vanquishing. All along I have argued that the notion of independence underlying this supposed antagonism must be undermined in the order of co-making. As we declared at the end of part two, consciousness is between us and the world and hence the human community itself arises there as well—namely, *society*, as Marx understood it, involving "the completed, essential unity of man with nature, the true resurrection of nature, the fulfilled naturalism of man and the humanism of nature" (1967:306). For Marx, in other words, a truly human science is simply a science that embraces the new sense of objectivity, which, again, extends to the world the community structure that contemporary philosophy and sociology (Restivo) have found at the origins of inquiry.

Without a truly human science, however, the very objects in the world, especially the *practico-inert* that embodies the collective *praxis* of human beings (Sartre, 1976), will be resolved as referring to the faces of the "experts" that are supposed to control knowledge about these objects. Marx was especially concerned to show that under capitalism, whatever objects workers make, the objects refer to the faces of the capitalists in general and to the "experts" that control the knowledge on the capitalists' behalf in particular. (As we already reminded ourselves in section three of this part, "experts" are read in the Soviet Union as well.) I am thinking here of such current objects as automobiles that are constructed so that their repair requires special tools, tools made for that model of automobile *alone*. We can recall Marx's words about the tools of manufacturing that became specialized to one task as opposed to another so that no movement could separate them: what separated them was the point of view of the capitalists or their surrogate "experts" who inspected the use of the tools. So, too, today do the special tools and the automobiles for which they are designed refer to the point of view of the capitalists or their surrogate "experts." The very props out of which we make our lives tend to be read so as to remove us from the circles that control the knowledge about them. Hence, we must live exactly as José in our lives, as people who were once literate—or at least able to be literate—and who were subsequently induced to become illiterate.

Again, another illiteracy pervades our society, one that need not necessarily be removed should we actually resolve the problem that Jonathan Kozol so eloquently described. This other illiteracy still has the same form as that of José, however, inasmuch as the problem turns on persons who do not have any sense of possessing the words in terms of which their lives hang in the balance. I am referring to a general *technical illiteracy* that induces a person to listen to an "expert" as if the technical words employed by the "expert" were automatically beyond that person's proper sphere of judgment. A "client" may be told, for example, that the proper judgment is really beyond one's comprehension so that one must simply trust "expert" judgment. But how does one trust when the very ground of trust is obscured by the invisibility of the governance at work here?

Throughout this section the practice of obscuring the grounds of trust has been described in terms of the way in which the reference to the faces of "experts" can be read into our bodies and into the objects in the world around us. I have chosen examples that cut across the major disciplines of inquiry—from natural to social science—and across the major categories of everyday life—from childhood to adulthood. At this juncture in our history it may well not need saying that the alienation of "expert" inspection has become profound: we have lost so much of our sense of trust, trust in ourselves and in the ways of the world, that many of us no longer hold any faith in the future. A "new politics of truth" based on sensuality and circles of consensus would certainly restore our missing trust and hence the ground of a new faith in the future: we must engage at all turns in the co-making of inquiry. A "new politics of truth" can be based, in other words, on the terms of posture.

In the introduction I noted that my own practice of the osmosis method is confined largely to young persons at a university, considerably older than Dennison's young persons. Nevertheless, for those who have complained to me over the years about being, in effect, under the spell of readings, readings that involve the current "discursive regime" of our society, the use of the osmosis method is always the crucial first step. As I have endeavored to show in this and the last sections, the osmosis method is just as crucial for their academic inquiries as it is for their everyday lives. Under the spell of readings, they cannot help but practice the "old politics of truth," forever struggling to discover the one correct reading that will resolve each of their problems. So, too, in general, in our pursuit of a better life, must we begin by breaking out of the spell of the readings that currently preoccupy us, for they constitute precisely the "discursive regime" that keeps us from that better life.

To take one example that cuts across academic inquiries and everyday affairs, and as well that continues the feminist theme of the examples above, research has established that it is a woman's lack of control of the informal, face-to-face struggle over how to read persons that essentially determines her lower chances for advancement in our institutions (Keller; Long; Lorber; Weisstein). Hence, more generally, the women's movement has not brought about a significant change in the structure of "political authority" in our society by petitioning that authority. Even though the resulting concessions of "political authority" have certainly been necessary, they have only appeared to be sufficient, especially in the case of the enfranchisement of women (Blau; Firestone; Freeman, Kaufman; Norgren; Shortridge).

While in his earlier work Foucault did not fully realize this character of "strategies of power" (as we noted, for example, at the beginning of section three of this part), he too came to stress the importance of the informal, face-to-face struggle over how to read persons:

> To say that "everything is political," is to affirm this ubiquity of relations of force and their immanence in a political field. . . . Generally speaking I think one needs to look . . . at how the great strategies of power encrust themselves and depend for their conditions on the level of micro-relations of power. (Foucault, 1980:189, 199)

This last comment drove one of his interviewers to paraphrase his position quite appropriately: "power as exercised from above is an illusion" (200). Or as Foucault himself put it (188), *"power is constructed": the governance in question is invisible.* It is precisely the spell of our current readings that constitutes the real ground of the "relations of force" in our society.

We cannot proceed directly, therefore, to form circles of consensus. Wherever we are still under the spell of our current readings, the osmosis method can break that spell, thereby creating additional openings for the practice we advocated in the last section in the spirit of Baldelli's "social anarchism": the removal of acts of inspection one by one. We aim gradually to retrieve the residue of our original posture, that is, the direct efficacy of movement and the associated solidarity of sensuality. As this process proceeds, the centralized structure of "political authority" will gradually give way to the decentralized structure of "natural authority," and on this basis we can in turn gradually replace the former, centralized structure with local circles of consensus. (Goodman thought of this process as "the extension of spheres of free action until they make up most of social life" (2).) Once again, although "experts" are invisible, expertise need not be: if expertise is visible and thereby constitutes "natural authority," then the necessary structure of activities will simply arise in the "natural movement," the "together-step," of the persons in question—hence the solidarity of sensuality. But if expertise is not visible, it can still flourish in local circles of consensus.

In the same vein, Paul Loeb recently advocated "human-scaled efforts" to achieve a better life, namely, "village politics":

> Complementing grand efforts like nationwide demonstrations or statewide referendums, these more intimate actions provide building blocks for broader change. They assert that power lies not only with the kingmakers in their insulated suites in Washington, D.C., but with ordinary citizens in the do-

> mains they inhabit. . . . [And] these local efforts can also push existing sources of information to challenge the false balm drummed in by official experts on the nightly news—push sources humble as church newsletters or Saturday bridge clubs. (80–2)

The aim is, in effect, to break the spell of our current readings, a spell that induces us to feel that we cannot do anything to change our lives for the better. Loeb would certainly applaud the idea of understanding our relationships with each other as potential, if not actual, circles of consensus.

I also help young persons at my university and elsewhere to practice "peer facilitation." Again, their relationships with each other are thought of as potential, if not actual, circles of consensus. They should not so much try to tell each other the truth, or their own answers to the crucial questions of their lives, but rather try to raise these questions with each other, keeping each other honest and looking as much as possible for the consensus that can only arise in this kind of context, indeed, a context of the new sense of objectivity developed in part two. To put it another way, peer facilitation aims to keep alive as much as possible the original consensus that actually grounds our conversations. Thereby we will be in a position not only to grasp our original free choices, but also to remake these choices under the new circumstances of our lives.

It is precisely our not being in this position in general in our lives that has caused us to react so slowly—perhaps too slowly—to our current problems. We have become much too tight about our readings of each other, and as a result we flee the very aleatory ground on which we could actually loosen up and remake our society. This ground is really so easy to realize as well, especially when it comes to retrieving the direct efficacy of our movement and the associated solidarity of sensuality. People who have tried the osmosis method have always reported that it gave them at least a glimpse of what it would mean to break the spell of their readings. The ground of aleatory inquiry is still in our bodies; we were not seduced by the posture of points of view long enough ago for that seduction to be wired into our bodies. The co-making of inquiry—the common sense of the human community—is just beneath the surface of the readings that induce us to look the other way, toward the "expert" inspectors of our lives.

We must establish communities of inquiry in which we work *with* each other and the world as our bodies make possible. Certainly we need to understand, and to bring under conscious control, our ability to conduct inquiry on our feet as we used to do before the dawn of the modern era. And when we get off our feet, we must always try to keep the uniquely human good from degenerating into a reading under the control of a hierarchy of points of view, that is, under "political authority" and the "old politics of truth." Inquiry begins between us and the world, and we must preserve its solidarity by forming circles of consensus so that we establish the uniquely human good based on the equality of all points of view precisely as the points of view of persons. In doing so we will automatically begin both to conduct what is now called "science for the people" and to respect nature in the spirit of the "new ecology" (Merchant).

Again, as I put it at the end of the previous section, the practice of sensual friendship will let loose the healing energies of our common sense. That this result can be defended in the terms of posture was the justification for writing this book in the first place. Instead of being under the spell of our current readings—as if, to call again on our most poignant example, Dennison were first and foremost to have addressed José's face apart from the order of movement of his body—we need to recognize the ways we *can* make places in the world. Indeed, we need to recognize that *and only that* in order to choreograph a better life, a dance of sensual solidarity and consensus, celebrating the order of co-making of inquiry. *All we need to do is to develop the options of human posture.*

Bibliography

Alpern, Mathew. "Eye Movements," in *Handbook of Sensory Physiology, vol. VII/4: Visual Psychophysics,* edited by D. Jameson and L. M. Hurvich, Berlin: Springer-Verlag, 1972.

Aries, Philippe. *Centuries of Childhood,* trans. Robert Baldick, New York: Random House, 1962.

______. *The Hour of Our Death,* trans. Helen Weaver, New York: Random House, 1982.

______. *Western Attitudes Toward Death,* trans. Patricia Ranum. Baltimore: Johns Hopkins University Press, 1981.

Aristotle. *Introduction to Aristotle,* edited by Richard McKeon, Modern Library. New York: Random House, 1947.

______. *Works, Volume Three,* edited by W. D. Ross, Oxford: Clarendon Press, 1955.

Aspect, Alain, J. Dalibard, and G. Roger. *Physical Review Letters, vol. 49,* 1982: 1804.

Baldelli, Giovanni. *Social Anarchism,* Chicago: Aldine-Atherton, 1971.

Bart, Pauline B. "Depression in Middle-Aged Women," in *Woman in Sexist Society,* edited by V. Gornick and B. K. Moran. New York: New American Library, 1972.

Bateson, Gregory. *Mind and Nature.* New York: Dutton, 1979.

Baumgardt, E. "Threshold Quantal Problems," in *Handbook of Sensory Physiology, vol. 7/4: Visual Psychophysics,* edited by D. Jameson and L. M. Hurvich, Berlin: Springer-Verlag, 1972.

Beattie, Geoffrey W., and Carol A. Beattie. "Postural Congruence in a Naturalistic Setting." *Semiotica, vol. 35,* 1981: 41.

Bell, J. S. *Physics, vol. 1,* 1964: 195.

Bentham, Jeremy. *Works, vol. 4,* edited by John Bowring. Edinburgh: 1843.

Berkeley, George. *Philosophical Writings,* edited by David Armstrong. New York: Collier, 1965.

______ . *Three Dialogues Between Hylas and Philonous,* edited by R. M. Adams. Indianapolis: Hackett, 1979.

Bernard, Jessie. "The Paradox of the Happy Marriage," in *Woman in Sexist Society,* edited by V. Gornick and B. K. Moran. New York: New American Library, 1972.

Blau, Francine D. "Women in the Labor Force: An Overview," in *Women: A Feminist Perspective,* edited by Jo Freeman. Palo Alto: Mayfield Publishing, 1984.

Bohm, David. *The Special Theory of Relativity.* New York: W. A. Benjamin, 1965.

______ . *Wholeness and the Implicate Order.* London: Routledge and Kegan Paul, 1980.

Bohm, David, and A. Baracca, B. J. Hiley, and A. E. G. Stuart. "On Some New Notions Concerning Locality and Nonlocality in the Quantum Theory," *Il Nuovo Cimento, vol. 28B,* 1975: 453.

Bohr, Niels. "Can Quantum-Mechanical Description of Physical Reality Be Considered Complete?" in *Physical Reality,* edited by Stephen Toulmin. New York: Harper and Row, 1970.

Boyd, Doug. *Rolling Thunder.* New York: Dell, 1974.

Braudel, Fernand. *The Structures of Everyday Life,* trans. Sian Reynolds, New York: Harper and Row, 1981.

Brucker, Gene ed. *People and Communities in the Modern World, vol. 1.* Illinois: Dorsey Press, 1979.

Burtt, Edwin Arthur. *The Metaphysical Foundations of Modern Physical Science.* Garden City, New York: Doubleday, 1954.

Capek, Milic. *The Philosophical Impact of Contemporary Physics.* New York: Van Nostrand, 1961.

Castaneda, Carlos. *The Eagle's Gift.* New York: Simon and Schuster, 1981.

Cazeneuve, Jean. *Lucien Levy-Bruhl,* tran. Peter Riviere. New York: Harper and Row, 1973.

Ceram, C. W. *Gods, Graves, and Scholars,* trans. E. B. Garside and Sophie Wilkins. New York: Bantam, 1980.

Chesler, Phyllis. "Patient and Patriarch: Women in the Psychotherapeutic Relationship," in *Woman in Sexist Society,* edited by V. Gornick and B. K. Moran. New York: New American Library, 1972.

______ . *Women and Madness.* New York: Avon, 1973.

Clauser, John, and Abner Shimony. "Bell's Theorem: Experimental Tests and Implications," *Reports on Progress in Physics, vol. 41,* 1978: 1881.

Crick, F. H. C. "Thinking about the Brain," *Scientific American* (September), 1979.

Cronon, William. *Changes in the Land.* New York: Hill and Wang, 1983.

Culler, Jonathan. "Jacques Derrida," in *Structuralism and Since: From Levi-Strauss to Derrida,* edited by John Sturrock. Oxford: Oxford University Press, 1981.

Dennison, George. *The Lives of Children.* New York: Random House, 1970.

Derrida, Jacques. *Positions,* trans. Alan Bass. Chicago: University of Chicago Press, 1982.

______. *Writing and Difference,* trans. Alan Bass. Chicago: University of Chicago Press, 1978.

Descartes, René. *Meditations on First Philosophy,* trans. Donald Cress. Indianapolis: Hackett, 1983.

Dijksterhuis, E. J. *The Mechanization of the World Picture: Pythagoras to Newton.* trans. C. Dikshoorn. Princeton: Princeton University Press, 1986.

Donzelot, Jacques. *The Policing of Families,* trans. Robert Hurley. New York: Random House, 1979.

Einstein, Albert. *Relativity: The Special and General Theory.* trans. Robert Lawson. New York: Crown, 1961.

Einstein, Albert, B. Podolsky and N. Rosen. "Can Quantum-Mechanical Description of Physical Reality Be Considered Complete?" in *Physical Reality,* edited by Stephen Toulmin. New York: Harper and Row, 1970.

Eliade, Mircea. *The Myth of the Eternal Return.* trans. Willard Trask. Princeton: Princeton University Press, 1974.

Estes, Caroline. "Consensus," *Social Anarchism, number 10,* 1985.

Evans, Cedric O., and John Fudjack. *Consciousness,* Baton Rouge, Louisiana, 1978.

Firestone, Shulamith. "On American Feminism," in *Woman in Sexist Society,* edited by V. Gornick and B. K. Moran. New York: New American Library, 1972.

Foucault, Michel. *Discipline and Punish,* trans. Alan Sheridan. New York: Random House, 1979.

______. *Power/Knowledge,* edited by Colin Gordon, New York: Random House, 1980.

Foucault, Michel, and Richard Sennett. "Sexuality and Solitude," New York, 1978.

Frederick, John. "Mythology and Culture: An Inquiry into the Nature of Language," Master's thesis, Rensselaer Polytechnic Institute, New York, 1986.

Freedman, Daniel, and Peter van Nieuwenhuizen. "The Hidden Dimensions of Spacetime." *Scientific American* (March), 1985.

Freeman, Jo. "The Women's Liberation Movement: Its Origins, Structure, Activities, and Ideas," in *Women: A Feminist Perspective,* edited by Jo Freeman. Palo Alto: Mayfield Publishing, 1984.

Freud, Sigmund. *Collected Papers, vol. 4,* trans. Joan Riviere. London: Hogarth, 1950.

______ . *General Introduction to Psychoanalysis,* trans. Joan Riviere. New York: Pocket Books, 1971.

______ . *Standard Edition of the Complete Psychological Works, vol. 6,* trans. James Strachey, with A. Freud, A. Strachey, and A. Tyson. London: Hogarth, 1960.

Frisch, Karl von. *Bees: Their Vision, Chemical Senses, and Language.* Ithaca, Cornell University Press, 1976.

______ . *The Dance Language and Orientation of Bees,* trans. L. E. Chadwick, Cambridge: Belknap/Harvard, 1967

Furley, David. "Aristotle and the Atomists on Motion in a Void." In *Motion and Time, Space and Matter,* edited by Peter K. Machamer and Robert G. Turnbull. Columbus: Ohio State University Press, 1976.

Gibson, James. *The Ecological Approach to Visual Perception.* Boston: Houghton Mifflin, 1979.

Gilligan, Carol. *In A Different Voice.* Cambridge: Harvard University Press, 1983.

Goffman, Erving. *Asylums,* Anchor. Garden City: Press, 1961.

______ . *Gender Advertisements.* New York: Harper and Row, 1979.

______ . *Relations in Public.* New York: Harper and Row, 1972.

Goodman, Paul. "Reflections on Drawing the Line," in *Drawing the Line: Political Essays of Paul Goodman,* edited by Taylor Stoehr. New York: Free Life Editions, 1977.

Gough, Kathleen. "The Origin of the Family," in *Toward An Anthropology of Women,* edited by Rayna Reiter. New York: Monthly Review Press, 1975.

Hall, A. Rupert. *From Galileo to Newton.* New York: Dover, 1981.

Harding, Sandra. "Is Gender a Variable in Conceptions of Rationality?" In *Beyond Domination: New Perspectives on Women and Philosophy,* edited by Carol C. Gould, Rowman and Allanheld. New Jersey: Totowa, 1983.

Harris, Marvin. *Cannibals and Kings: The Origins of Cultures.* New York: Random House, 1977.

______ . *Cows, Pigs, Wars, and Witches: The Riddles of Culture.* New York: Random House, 1974.

Homer. *The Iliad,* trans. Robert Fitzgerald. Garden City: Anchor, 1975.

Honegger, Barbara. "A Shamanistic Seed-Dream Interpretation: The Bird-Man Mural at Lascaux."*Phoenix, vol. II/2,* 1978: 5.

Hubel, David. "The Brain," *Scientific American* (September), 1979.

Hubel, David, and Torsten Wiesel. "Brain Mechanism of Vision," *Scientific American* (September), 1979.

Hume, David. *An Enquiry Concerning Human Understanding,* in *The Empiricists.* Garden City: Anchor, 1974.

Illich, Ivan. *Toward a History of Needs.* Berkeley: Heyday Books, 1978.

Jammer, Max. *The Philosophy of Quantum Mechanics.* New York: John Wiley and Sons, 1974.

Jastrow, Robert. *The Enchanted Loom.* New York: Simon and Schuster, 1981.

Jaynes, Julian. *The Origin of Consciousness in the Breakdown of the Bicameral Mind.* Boston: Houghton Mifflin, 1976.

Johanson, Donald, and Maitland Edey. *Lucy: The Beginnings of Mankind.* New York: Warner Books, 1981.

Kant, Immanuel. *Critique of Pure Reason,* trans. N. K. Smith. New York: St. Martin's Press, 1965.

______. *Prolegomena to Any Future Metaphysics,* trans. Paul Carus with revisions by James Ellington. Indianapolis: Hackett, 1977.

Kaufman, Debra Renee. "Professional Women: How Real Are the Recent Gains?" In *Women: A Feminist Perspective,* edit. Jo Freeman. Palo Alto: Mayfield Publishing, 1984.

Keller, Evelyn Fox. *Reflections on Gender and Science.* New Haven: Yale University Press, 1985.

Kozol, Jonathan. *Illiterate America.* New York: New American Library, 1985.

Kripke, Saul A. *Wittgenstein on Rules and Private Language.* Cambridge: Harvard University Press, 1982.

Krupp, E. C. *Echoes of the Ancient Skies.* New York: Harper and Row, 1983.

Kuhn, Thomas S. *The Structure of Scientific Revolutions.* Chicago: University of Chicago Press, 1970.

Laing, R. D. *The Divided Self.* Baltimore: Penquin, 1971a.

______. *The Politics of Experience.* New York: Ballantine, 1970.

______. *The Politics of the Family.* New York: Random House, 1971b.

Laing, R. D., and Aaron Esterson. *Sanity, Madness, and the Family.* Baltimore, Penquin, 1970.

Leacocke, Eleanor. *Myths of Male Dominance.* New York: Monthly Review Press, 1981.

Leakey, Richard, and Roger Lewin. *Origins.* New York: Dutton, 1982.

Leibniz, Gottfried W. F. V. *The Monadology,* trans. George Montgomery, with revisions by Albert Chandler. In *The Rationalists.* Garden City: Anchor, 1974.

Lindberg, David. *The Theories of Vision from Al-Kindi to Kepler.* Chicago: University of Chicago Press, 1981.

Linhart, Robert. *The Assembly Line,* trans. Margaret Crosland. Amherst: University of Massachusetts Press, 1981.

Loeb, Paul Rogat. *Hope in Hard Times: America's Peace Movement and the Reagan Era.* Lexington: Lexington Books, 1987.

Long, J. Scott. "The Origins of Gender Differences in Science: The Initiation of Cumulative Disadvantage," GTE Lecture, Department of Science and Technology Studies, Rensselaer Polytechnic Institute, New York, 1987.

Lorber, Judith. "Trust, Loyalty, and the Place of Women in the Informal Organization of Work." In *Women: A Feminist Perspective,* ed. Jo Freeman. Palo Alto: Mayfield Publishing, 1984.

Lukes, Steven. "Relativism in its Place." In *Rationality and Relativism,* ed. M. Hollis and S. Lukes. Oxford: Blackwell, 1982.

Marx, Karl. *Writings of the Young Marx on Philosophy and Society,* trans. and ed. Loyd Easton and Kurt Guddat. Garden City: Anchor, 1967.

______. *Capital, Volume One,* trans. Ben Fowkes. New York: Random House, 1977 .

Matin, Leonard. "Eye Movements and Perceived Visual Direction." In *Handbook of Sensory Physiology, vol. VII/4: Visual Psychophysics,* ed. D. Jameson and L. M. Hurvich. Berlin: Springer-Verlag, 1972.

Merchant, Carolyn. *The Death of Nature: Women, Ecology, and the Scientific Revolution.* New York: Harper and Row, 1983.

Merleau-Ponty, Maurice. *Phenomenology of Perception,* trans. Colin Smith. London: Routledge and Kegan Paul, 1962.

______. *The Primacy of Perception,* ed. James Edie. Illinois: Northwestern University Press, 1964a.

______. *Signs,* trans. Richard McCleary. Illinois: Northwestern University Press, 1964b.

Morris, Desmond, Peter Collett, Peter Marsh, and Marie O'Shaughnessy. *Gestures: Their Origins and Distribution.* New York: Stein and Day, 1980.

Musashi, Miyamoto. *A Book of Five Rings.* Woodstock: Overlook Press, 1974.

Neihardt, John. *Black Elk Speaks.* New York: Washington Square Press, 1972.

Neumann, Erich. *The Great Mother,* trans. Ralph Manheim. Princeton: Princeton University Press, 1973.

______ . *The Origins and History of Consciousness,* trans. R. F. C. Hull. Princeton: Princeton University Press, 1973.

Newton, Isaac. *Newton's Philosophy of Nature: Selections From His Writings,* ed. H.S. Thayer. New York: Hafner, 1974.

Nietzsche, Friedrich. *On the Genealogy of Morals,* trans. Walter Kaufman and R. J. Hollingdale. New York: Random House, 1969.

Norgren, Jill. "Child Care." In *Women: A Feminist Perspective,* ed. Jo Freeman. Palo Alto: Mayfield Publishing, 1984.

Ornstein, Robert. *The Psychology of Consciousness.* New York: Harcourt, Brace, and Jovanovich, 1977.

Pearse, Innes H., and Lucy H. Crocker. *The Peckham Experiment.* New Haven: Yale University Press, 1946.

Pinxton, Rix, Ingrid van Dooren and Frank Harvey. *The Anthropology of Space.* Philadelphia: University of Pennsylvania Press, 1983.

Plato. *The Phaedrus,* trans. W. C. Helmbold and W. G. Rabinowitz. Indianapolis: Bobbs-Merrill, 1983.

______ . *The Theaetetus,* trans. Francis Cornford. In *Plato's Theory of Knowledge.* Indianapolis: Bobbs-Merrill, 1957.

______ . *The Timaeus,* trans. Francis Cornford. In *Plato's Timaeus.* Indianapolis: Bobbs-Merrill, 1959.

Postman, Neil. *The Disappearance of Childhood.* New York: Dell, 1984.

Pribram, Karl. *Languages of the Brain.* Englewood Cliffs: Prentice-Hall, 1971.

Quine, Willard Van Orman. *Word and Object.* Cambridge: MIT Press, 1960.

Reichenbach, Hans. *From Copernicus to Einstein.* New York: Dover, 1980.

Resnick, Robert, and David Halliday. *Physics.* New York: John Wiley and Sons, 1977.

Restivo, Sal. "Modern Science as a Social Problem." *Social Problems, vol. 35/3* (June), 1988: 206.

Rorty, Richard. *Philosophy and the Mirror of Nature.* Princeton: Princeton University Press, 1980.

Rubin, Melvin, and Gordon Walls. *Fundamentals of Visual Science.* Illinois: Charles C. Thomas, 1972.

Sartre, Jean-Paul. *Being and Nothingness,* trans. Hazel Barnes, New York: Washington Square Press, 1973.

______. *Critique of Dialectical Reason,* trans. Alan Sheridan-Smith, ed. Jonathan Ree, London: New Left Books, 1976.

Scheff, Thomas. *Being Mentally Ill.* New York: Aldine, 1984.

Sennett, Richard. *The Fall of Public Man: On the Social Psychology of Capitalism.* New York: Random House, 1978.

Shimony, Abner. "Hidden Variables Theories, Bell's Theorem, and Locality." Paper presented at SUNY-Albany Conference on Fundamental Questions in Quantum Mechanics, April, 1984.

______. "Implication's of Bell's Theorem." Paper presented at SUNY-Albany Physics Colloquium, October, 1982.

Shortridge, Kathleen. "Poverty is a Woman's Problem." In *Women: A Feminist Perspective,* ed. Jo Freeman. Palo Alto: Mayfield Publishing, 1984.

Skinner, B. F. *Science and Human Behavior.* New York: Free Press, 1965.

Smith, Morton. *Jesus the Magician.* New York: Harper and Row, 1978.

Spinoza, Benedict De. *The Ethics,* trans. R. H. M. Elwes. In *Works of Spinoza, vol. 2,* New York: Dover, 1955.

Stanford, W. B. *The Sound of Greek.* Berkeley: University of California Press, 1967.

Steil, Janice M. "Marital Relationships and Mental Health: The Psychic Costs of Inequality." In *Women: A Feminist Perspective,* ed. Jo Freeman. Palo Alto, Mayfield Publishing, 1984.

Strawson, P. F. *Individuals.* Garden City: Anchor, 1963.

Suzuki, Shunryu. *Zen Mind, Beginner's Mind.* New York: Weatherhill, 1983.

Szasz, Thomas. *The Manufacture of Madness.* New York: Dell, 1970.

Thass-Thienemann, Theodore. *The Interpretation of Language, vol. 1.* New York: Jason Aronson, 1973.

Thompson, William Irwin. *The Time Falling Bodies Take to Light: Mythology, Sexuality, and the Origins of Culture.* New York: St. Martin's Press, 1981.

Tobias, Phillip. *The Brain in Hominid Evolution.* New York: Columbia University Press, 1971.

Vico, Giambattista. *The Autobiography,* trans. Max Fisch and Thomas Bergin. Ithaca, Cornell University Press, 1983.

______. *The New Science,* trans. Thomas Bergin and Max Fisch. Ithaca: Cornell University Press, 1984.

______. *Selected Writings,* ed. trans. Leon Pompa. Cambridge: Cambridge University Press, 1982.

Ward, Colin. *Anarchy in Action,* New York: Harper and Row, 1974.

Weisstein, Naomi. "Psychology Constructs the Female." In *Woman in Sexist Society,* ed. V. Gornick and B. K. Moran. New York: New American Library, 1972.

Whorf, Benjamin Lee. *Language, Thought, and Reality.* Ed. J. B. Carroll. Cambridge: MIT Press, 1984.

Wittgenstein, Ludwig. *Philosophical Investigations.* Trans. G. E. M. Anscombe. New York: Macmillan, 1968.

———. *Tractatus Logico-Philosophicus.* Trans. D. F. Pears and B. F. McGuinness. London: Routledge and Kegan Paul, 1972.

Zukav, Gary. *The Dancing Wu Li Masters: An Overview of the New Physics.* New York: Bantam, 1984.

Index